African catfish farmer's handbook

Willy Fleuren

Published in collaboration with Starcrest Integrated Farms Ltd. and S.C.C. Ltd.

STARCREST INTEGRATED FARMS LIMITED

SCC

African catfish farmer's handbook

African catfish farmer's handbook

Willy Fleuren

Published by Willy Fleuren in collaboration with Starcrest Integrated Farms Ltd. and S.C.C. Ltd.
Abuja, Nigeria 2019

Layout cover and inside work: Print Service Ede
Photos: Gerrit Fleuren Ventures
Printed and bound: Print Service Ede

Source references of illustrations

p. 12, 29, 32, 40, 43, 48, 60, 63, 66, 88, 99, 100, 107, 196, 222: Gerrit Fleuren Ventures
p. 23, 24: Practical manual for the culture of the African catfish (Clarias gariepinus) (1986).W.J.A.R Viveen, C.J.J Richter, P.G.W.J. van Oordt, J.A.L. Janssen et al.
p. 105: Muhammad Umar Aziz, student at Bahauddin Zakariya University
p. 123: World Climate Guide
p. 148: The nature of fish diseases (2002). J.W. Warren, U.S. Department of the Interior Fish and Wildlife Service La Crosse, WI
p. 150: Cornell Composting
p. 152: JBL Testlab
p. 154: Fondriest Environmental Products

ISBN 9789083013732

TABLE OF CONTENTS

FOREWORD

Today's world of fish farming, filled with lots of theories of so-called experts, challenges our success in fish farming.

We need to remember that some people in this world are doing the right thing, by getting the required amount of water which is the most important factor, the required quantity and quality of fish feed and other essentials for farming, and getting their results.
One of the main subjects that Willy Fleuren, my boss and mentor, loves to say is, "Do the right things, stop cutting corners."

This book you are holding is the result of Willy Fleuren's long time experience in building and installing hatcheries, grow-out systems, and fish farm management in both Africa and other parts of the world.

As I travelled and worked with him, both in Nigeria and other countries of Africa, over the years I saw the benefits of doing the right thing in fish farming.
This legacy edition of Willy Fleuren's book, filled with special added contents, is a valuable gift for the present and the future generation of fish farmers and also for students of aquaculture and fisheries in Africa and beyond, to answer their hunger and desire for the right practical techniques of profitable fish farming and management.

By Wale Olabode, Aquagarden,
Fish Farm,
FCT Abuja - Nigeria

1

INTRODUCTION - REASONS WHY FISH FARMS DON'T SUCCEED

1. INTRODUCTION OF AQUACULTURE IN NIGERIA
2. REASONS WHY FISH FARMS DON'T SUCCEED

1.1 INTRODUCTION OF AQUACULTURE IN NIGERIA

Aquaculture is the production of aquatic organisms such as shrimps, fish, shellfish and seaweeds. In Nigeria however, aquaculture is almost limited to fish farming only. The species which are suitable to farm in Nigeria, are limited too. Depending on the biology of the species, market, availability of seed and suitable feeds, I consider three species to be feasible for farming in Nigeria: African catfish (Clarias gariepinus), Tilapia (Oreochromis niloticus) and Carp (Cyprinus carpio). From these three species the vast majority of the production is of African catfish, also called the walking catfish or barbel, the local name in South Africa. In this handbook I only focus on the production of African catfish, because it has a wide acceptance in Nigeria, it grows very quickly and it can be economically farmed in small scale enterprises with relatively low investments. On the African continent fish farming is not widespread. There are only two countries which really have developed fish farming: Egypt and Nigeria. Egypt is mainly focusing on tilapia production and Nigeria is mainly focusing on African catfish production.
In the early days after the Nigerian independence, 'fish farming' and 'fishing' were mixed up terms, but they have a totally different meaning. Fish farming is done in a controlled water body, fish are stocked and fed deliberately. After a growing period, the fish are harvested in a controlled manner to easily crop the fish. Fishing is catching fish in natural water bodies with seine nets, fish traps, angling and other fishing equipment. What is caught is always a surprise. The term 'fisheries' is actually covering both activities.

In the early seventies the first pioneers started with real fish farming, driven by a growing population and increased prices for fish. They however faced a serious problem: the know-how for fish farming was not available and good quality feeds and seeds were very scarce.
In the eighties fish farming started to change through the introduction of new farming techniques, which resulted in increased fingerling production of catfish, and the introduction of imported starter feed.

In the nineties fish farming really started to boom and the RAS (recirculation aqua culture system) was introduced. This system is highly intensive and does not need much space. You actually can farm successfully in the middle of town. The awareness of fish farming increased exponentially and already a thousand fish farmers had started their own farm. However, in the rush of starting, many farmers stopped their farming activities as fast as they had started them and lost a lot of money. Till today we see many newcomers, and still many of them end up disappointed and leave this volatile agribusiness. This handbook is hopefully a tool for anybody who wants to start a fish farm, and an inspiration to gain the knowledge to get started on a solid foundation, so there will be no drop out.

That is why this book is full of examples, formats and calculations. I realise that the prices of materials, fish feeds, fish, energy and other costs will fluctuate in time and location. That doesn't mean that the book will be outdated in the years to come, because the formulas used in this book are universal. You can just fill in the current prices to get a customized outcome and draw actual conclusions.

After the millennium the government positively interfered in the Agric sector. Agric loans were developed with single digit interest rates. However only big companies and companies which are well connected to the banks, were benefiting from those loans. Also, imported agric equipment, including for fish farming, can be imported duty-free. There are restrictions on the import of frozen fish and importers cannot acquire CBN funds for importation of frozen fish. The import of frozen tilapia and frozen catfish is completely forbidden, however frozen tilapia is still dripping in via The Republic of Benin. All this has stimulated the growth of fish farming in Nigeria. Nowadays frozen fish is very expensive.
The present administration takes diversification of the economy very seriously and is concentrating on the agric sector being a sector that can absorb a lot of unemployed youth and increase the sustainability of the food supply for the untamed growing population.
The government has created a good platform for the fish industry and multinationals have come to Nigeria to invest heavily in commercial state-of-the-art feed mills.

Catfish farming can be carried out in almost any state of the country. Claiming this, the production of catfish is actually primarily based in the south west of Nigeria (Figure 1). The Northern states have less water available, whereas some states, such as Taraba State (highland) and Plateau State, have enough water, but the climate is too cold for profitable catfish farming. The east and south-south are provided with good opportunities to farm fish but are not so developed yet.

MAP RANKING STATES IN NIGERIA WITH HIGHEST PRODUCTION OF CATFISH

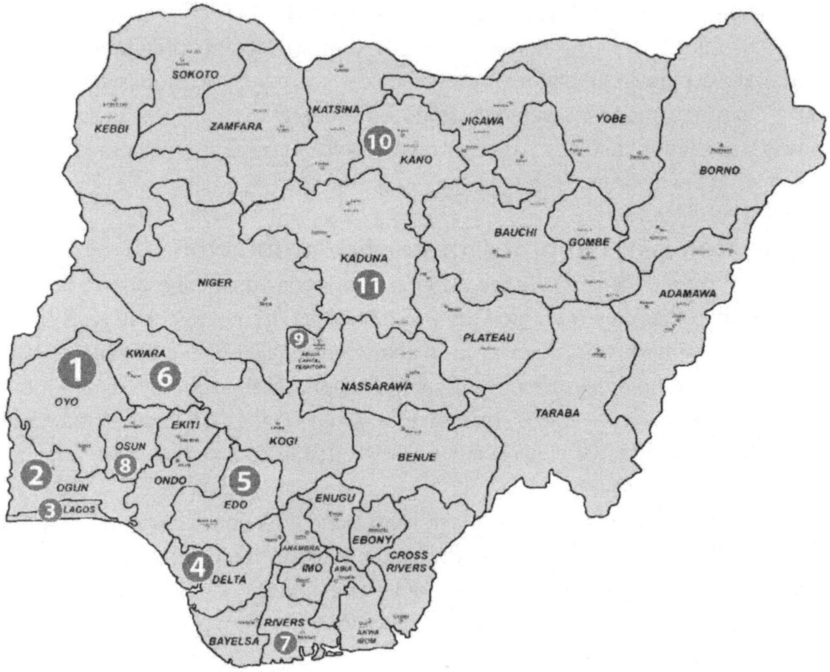

Figure 1. The ranking of states based on the level of catfish production.

Nigeria is blessed with a lot of water bodies: 264 medium and large dams with a combined storage capacity of 33 billion cubic metres of water. The dams can be used for pumping water into the fish farms or for direct farming by putting cages in the dam.

The demand for fish in Nigeria is around 7 kg/capita per year, but the recommended rate is 12 kg/capita per year. Fish production in Nigeria is hovering around 200,000 tons; the domestic catch is 450,000 tons, the remaining fish, which forms the bulk, is still imported.

1.2 REASONS WHY FISH FARMS DON'T SUCCEED

First of all, fish farms shouldn't fail because there is no market for catfish. The market for catfish is big and still growing, although in some areas they produce a lot of catfish but they don't eat as much. So, they are marketing it to other areas in the country. For example, in the Ibadan region a lot of fish is produced and transported to Abuja and Lagos by road. Marketing is part of the business and if totally neglected, it will result in very low sales prices. This makes it difficult to make a profit. Therefore, fish farms should have a regular supply of fish to be able to satisfy the customer on a regular basis. However, some farms sell their produced fish only two or three times a year. In this way you cannot build any customer relationship! In general, small producers have more problems selling their fish than the big producers.

The location is not appropriate for fish production

Many fish farmers underestimate the water consumption. Their conception is to fill the tanks and keep them full from growth cycle till harvest. It's not like that; read chapter 3 for more information. In order to keep the water quality within acceptable levels, you have to flush on a daily basis, preferably 24/7 and even increase it as the fish are gaining weight. Using boreholes for grow-out, which is the last stage of the growth cycle, is in many areas of Nigeria not really profitable for medium and large-scale production.
A borehole can work only on a small scale, and the production is limited to only 1000 kg per month. Of course, there are always exceptions. In the region of Benin City, you will see good working boreholes with very large capacity (20 - 30 m^3 per hour) and clean water. Normally the supply of water should come from a river, stream or dam with availability of water throughout the year.

Some farms are based so far away from civilisation, that it is a challenge to get there. The roads are very bad, security is almost absent and no infrastructure is available. Customers will be discouraged to come and buy fish in such an area.

Some fish farms are not equipped with any way to drain water out of the farm. This is the same misconception as not having any water to flush, because of the perception that the ponds must be kept full. Solutions to drain the water into the soil cannot work because soon you will have a lagoon on your land. If you by all means want to drain onto your property, you will need 1000 m^2 for each ton/annum of production. For a production of 50 tons per annum, this is already 5 ha of land.

If your farm is located far from residential areas, you will have to provide good accommodation for key staff. If you do not, the staff will not be comfortable to work, and you won't get good staff with dedication and motivation. Most investors don't want to invest in accommodation because it's not related to production. Accommodation is something which will be implemented if profit is made. But for staff to suffer and stay 24/7 on the farm is not realistic. You will end up having no results at all.

Lack of knowledge, bad consultancy

In the land of the blind, the one-eyed man is king. So, I was invited by an investor to visit his farm in Nasarawa State (Nigeria). After a lot of work to build his fish farm, he was starting to feel uncomfortable about the set up. Indeed, he should feel uncomfortable because when I arrived at his farm, I was shocked about what had been done so far. To describe the situation of what I saw there; four holes in the ground, each with a mega size of around 3000 m^2 and at least four metres deep. No inlets, no outlets, no drainage system, no water supply. "But we have drilled two boreholes and I was told they can supply the ponds!", the investor protested. I told the investor: "Even when you run ten boreholes, it will not fill any of your holes in two months' time, neither in the rainy season nor in the dry season. The holes don't keep water because the soil is not holding water! It's simple as the ABC." "So, what is the way forward," he asked me. I told him to fill the holes and rethink the farm, but now with a professional and realistic approach.

Many farmers start a fish farm without the basic knowledge of fish farming. They are relying on so-called fish consultants who in rare cases are independent and rarely give you honest advice. This handbook should prevent you from making basic mistakes. Most farm owners are actually investors. People around the investor could have been talking about how good and profitable fish farming is, but he never went into the matter in detail. The fact that your neighbour is farming fish and is telling you how well the business is running, should not be the reason to start yourself. Investigate the business: examine the value chain, the risks involved, try to visit farms which are producing a substantial amount of fish and talk with successful farmers.

Successful farmers tend to be more honest than unsuccessful farmers because the latter won't like to tell you that they failed. If farmers fail, they always put the blame on the staff, the bad juveniles they have received, bad purchase of feed and bad advice. What about themselves? Was it not them who built the farm, who paid for the feed, who employed the staff and paid for the juveniles?

The investors who succeed in the business of fish farming, are dedicated, love the challenge and really go for it. They are well informed because they have gone to seminars, workshops, markets and farms before they start their own fish farm. They choose the location where to farm based on a SWOT analysis, not on sentiment. A sentimental reason could be a wish to start fish farming in your village where you hail from, because you want to do something for your family or the community.

Often juveniles or fingerlings are sold by consultants. They want you to buy a lot, even if you don't have the money, space or water to facilitate them. At first there is no problem, because the fish are still small. But when your consultant has moved on and the fish have grown bigger, you'll find yourself in trouble.
If you want to start a fish farm to gain experience, by all means start small.

Facility is poorly built and poorly set up

Many fish farms are built with inferior materials at the lowest cost. The perception is that fish farming is just a hole with water, either dug in the ground or built with blocks. Fish farming is not a low-class business. To keep the water in a pond or a concrete tank, means good construction, compacting, reinforced concrete, pillars and good plastering with correct mixtures of cement and sand, contracting good technicians with good tools. Very often ponds are made on a slope and the water level in the pond or tank varies according to the slope. Fish ponds should be built level and could have a slope on the bottom but not on the top. Mostly the pipes bought and used to supply water and discharge water are of such poor quality that it will result in leakages. Especially having inferior quality pipes and fittings under the concrete, is very bad, because repairing leakages in that spot is very difficult. At most farms the pipes to supply or to drain are too small also. It is again the perception that the fish pond doesn't need much water, only to fill. Drainage and supply have no priority. Always use high pressure pipes of 10 or 16 bar.

If ponds are built well, they will last for decades and eventually will always be cheaper.
All the pipes that are laid in the soil should be stabilised with sharp sand or stone dusk. This is to avoid the pipes breaking when people or machines go over them. Mostly, the pipes are laid just a few inches below the topsoil. It is just a matter of time before they start to leak. Always make a drawing of where all the pipes are situated, so that if you want to make adjustments or amendments in the future, you know exactly where the pipes are.

When you start setting up the farm, make a complete blueprint; get a survey of the land so that you become aware of the levels of the land and build according to the levels available, in order to make drainage and pumping as efficient as possible. Make a plan for the start-up stage, keeping in mind possible expansion in the future. Most of the time, farmers start building without any vision. This will make the farm look scattered and disorganised and running the farm will not be efficient.

Have in mind your storage facility, the ease of keeping your feed and materials and how to hand it out efficiently to the farm. The security facility should be situated at a central point in order to easily oversee the complete farm. Other facilities have to be situated in the right place also. In chapter 6 I give you an example of setting up a complete farm with all the facilities involved.

There is no supervision and the recruitment of staff is not taken seriously

The farm is ready to launch; you start employing people with fish farm qualifications and now it's time for you to relax and wait for the results to come. Unfortunately, it doesn't work this way. The farm needs intensive supervision. You have to show the staff that the farm is very important to you and that you are eager to get and see results. Supervision doesn't mean to put fear into the staff when things are not going as they should. Supervision means guiding and coaching the staff day by day, encouraging them to do the work with passion and helping them to solve the problems on the way instead of blaming them for mistakes. If the farm is managed by fear, the staff will start to hide all the mistakes happening on the farm and they will do eye service. It divides staff members and blame is pointed in all directions instead of putting heads together to solve problems. If a farm is running smoothly and production is ongoing, it's a mutual effort and not a one-man's achievement. Teamwork is the key to success at your farm. Pay your staff well, don't turn them into beggars.

Another mistake many farmers make is to rush things as soon as the farm is ready to start. The investment has to be paid back as soon as possible and the investor wants to see fish in all the ponds from day one. It doesn't work this way; first you start to run the facility without fish for a couple of days to be sure that the entire infrastructure is working fine boreholes, pumps, generator, solar etcetera. The brood stock is already conditioned and doing fine. If you start with juveniles, you start up with one or two ponds, not with all the ponds at the same time. You have a production plan and you don't deviate from it.

When you employ staff for the farm, it's not necessary to look for very experienced fish farmers, instead look at skills such as willingness and ability to learn, honesty, ability to work in a team and having a passion for the job. If you can find these kinds of people, you certainly will enjoy them as long as you supervise them and give them incentives when the farm is making progress. Try to find out about the background of the staff you are employing, get references and guarantors, home address etcetera. The staff you recruited, should be aware of the rules and the regulations on the farm. They should be aware that any issue of theft is taken to the police. Be careful not to give them a very narrow job description. A farm cannot operate if the staff are not cooperative in all the farm issues involved, because the job description gives them excuses for not doing things.

Lack of funding

Things will go wrong if the funding is not sufficient to complete the investments and/or the working capital runs out before the first sales. The investor/owner will start to cut corners and will be pressurized to get quick results. He will also put a lot of pressure on his staff; demanding results which are unrealistic. Quick results are usually very bad results and more misery. It can kill the farm from the beginning. Build a farm you can afford and have a realistic cash flow programme to back up what you want to do.

Theft

Some years ago, I was called by a farmer who was using the feed I was selling to him. He was very disappointed in the performance of the feed; the fish were not growing. I went to visit the farm and couldn't trace the problem. It seemed that the feed was not stolen and the fish were not stolen either, since the records and the sales were showing all the results and the stocked fish were retrieved at harvest minus the recorded mortality. The fish were just very small but didn't look malnourished. I left the farm and felt bad for him, but I knew we were overlooking something. Two months later he called me and told me what was going on. The manager of the farm was taking out the big fish in the evening but was smart enough to replace them with smaller fish to keep up with the number of fish which was guarded so anxiously. He apologized for accusing the feed being so bad and told me: "If Nigerians use their smart brains in a positive way and not for 419, Nigeria would be a better place!"

Unfortunately, theft is a big issue in many farms and it starts from lack of supervision and lack of good management. Most likely there are no standard operational procedures (SOP) and everybody in the farm does what he likes. In this type of environment, it's very tempting to start stealing because nobody knows what exactly is going on at the farm. Theft will not be spotted quickly. To be honest, even if you give your staff good salaries, it will not avoid any theft on the long term.

Commitment and passion to do a good job are the virtues you have to create in your staff. Give them the feeling that it's their own farm and that the destiny of the farm and their future is in their own hands. An environment where you start shouting at staff and threaten to sack them if they don't perform, will certainly result in stealing. Their minds are already out of the farm. As the manager you have to create a team spirit in your staff; in this way anybody who wants to do harm to the farm will be spotted and corrected before theft starts. If the staff are set up against each other, it will certainly lead to bad results. Be sure that the people you bring together can also work together as a team.

Some owners like to put people together who actually do not match; they believe the workers will control each other and it will reduce or avoid theft. I don't think this is the right way, it will lead to bad results. People who don't like each other will always like to see the other person's downfall. That means bad results at the farm, which will be blamed on one person only. Before you know it, the staff start blaming each other, instead of putting their heads together to solve problems and get results.

Theft within the farm:

- Stealing fish directly for own use or to sell.
- Selling fish for a low price and get a pay back from the buyer.
- Selling fish but giving the customer more than what he bought for it and getting a pay back in return.
- Buying feed or products with a commission from the supplier as a pay back.
- Selling fish farm inputs such as diesel, feed etcetera.
- Putting extra high prices for any maintenance to be done at the farm.
- The owner pays for the feed directly to the shop but the farm manager will not take all the feed to the farm. Instead he has an arrangement with the shop owner to leave some feed behind and get money in return. Or get some empty feed bags of that brand and fill it with an inferior feed in order to have a complete delivery.

Some of the theft can be a one-man activity but stealing fish or other products from the farm, has to be done in collaboration with other people which in most cases is the Security. The worst thing happens if you are employing workers who happen to be your family and they are the ones stealing at the farm. If you employ family members, please employ them on qualifications and integrity not on sentiment. Never employ people you cannot sack.

Fish farming is not a retirement job

For many people this will not be the information they want to read.

Many fish farmers, who start their business, have a retirement background. The main reason for that is that they have the money to start and plan ahead to start a fish farm the moment they are retired, having already invested in a plot of land some years before. Most of these retirees had a job as civil servant or office worker. Their work experience doesn't really fit the qualifications to be a fish farmer. Fish farming is a way of life and you need to be very committed in order to get profitable results. Neglecting the farm for a couple of days can give a drawback of many months. Most retirees rely on the people they hire, but you have to be a front soldier yourself in order to push your staff to be a front soldier too. After some time, it's getting worse; the drive has vanished and the fish farmers are not coming to the farm as frequently as before. The road is bad and other obligations are coming in. The farm will be visited only once a week accompanied by some phone calls to find out what is going on at the farm. Now the farm is managed and supervised by a weekend-farmer or even worse by a telephone-farmer. This farm is doomed to die and will never make any money. The staff are frustrated, no investments are made anymore and the farm, once a-dream-come-true, has become a painful ulcer and a drain of money.

2

THE BIOLOGY OF THE AFRICAN SHARPTOOTH CATFISH

1. THE SPECIES OF CATFISH AND THEIR GEOGRAPHICAL DISTRIBUTION
2. BIOLOGICAL DESCRIPTION
3. FISH ARE COLD-BLOODED ANIMALS

2.1 THE SPECIES OF CATFISH AND THEIR GEOGRAPHICAL DISTRIBUTION

In nature there are over 3000 species of catfish and you can find them on almost every continent. The size of catfish species can vary from less than 1 cm to over 2 m long. What all catfish have in common is the absence of scales and the presence of barbels, although they differ in number and size. Some of the catfish species are cultured and are important food fish for human beings, for example the African sharptooth catfish (Clarias gariepinus), the European catfish (Silurus glanis) and the American catfish (Ictalurus punctatus). Other catfish species are very beautiful, colourful and greatly appreciated as ornamental fish in aquaria and ponds.

The African sharptooth catfish belongs to the family of Claridae and has its geographical distribution over the whole continent of Africa. Because of their importance as food fish, African sharptooth catfish are exported to Asia as well as to South America and Europe where they are cultured and grown for human consumption. Their natural habitats are tropical swamps, lakes and rivers, preferably slow-moving water. If the water body dries up, they can even survive in the mud until the next rainy season.

2.2 BIOLOGICAL DESCRIPTION

Skin

African sharptooth catfish have a scaleless slimy skin, which is darkly pigmented, brown and black on the dorsal and lateral (top and side) parts of the body. The skin over the abdomen (belly) is lighter coloured grey to white. Depending on the colour of the water, the pigmentation changes. Catfish tend to adapt to the colour of the water, in order to be less visible. The slime of the catfish acts as a defence mechanism against parasites and bacteria. Always be careful not to damage or reduce the slime of the skin, otherwise you will certainly get problems which can cause fish mortality.

Mouth

With its wide mouth, the African sharptooth catfish has the ability to feed on a variety of food types, ranging from small zooplankton to fish. It is able to suck benthos, for example mosquito larvae from the bottom, tear pieces of cadavers with its small teeth on its jaws and swallow prey, such as whole fish. The mouth circumference of this gape-limited predator, which is about ¼ of its total length, determines the maximum size of its prey. An African sharptooth catfish of 30 cm long (approximately 200 g) has a mouth circumference of about 7.5 cm. This enables it to encompass the body circumference of

small tilapia of up to 8 - 10 cm, which makes the African sharptooth catfish an excellent predator for controlling overpopulation of tilapia in ponds.
Catfish are cannibalistic by nature and start to eat each other from the first day of feeding onwards.

Barbels, olfactory organs and eyes

Around the mouth of the African sharptooth catfish eight barbels can be distinguished (Figure 2). The barbels can be moved without moving the mouth itself. Close to the nasal barbels two olfactory organs are located (Figure 2). Catfish recognise their prey mainly by touch and smell. This is of relevance during feeding at night and in highly turbid or muddy waters. Catfish have small eyes and don't rely very much on their eyesight.

Fins

The median fins of the African sharptooth catfish consist of a dorsal, a caudal and an anal fin, while the paired fins consist of the pectoral and ventral fins (Figure 3a). The pectoral fins have developed strong spines (Figure 3b) which have a locomotory and protective function. The fish are capable of migrating over land by sculling with their tail, as they elbow along on their spines. The spines of the African sharptooth catfish are not poisonous, whereas other catfish species do have poisonous spines.
The pectoral fins come in very handy when taking the fish out of a net or bowl, because the moment you touch the catfish, they will open their fins as a reflex to defend themselves.

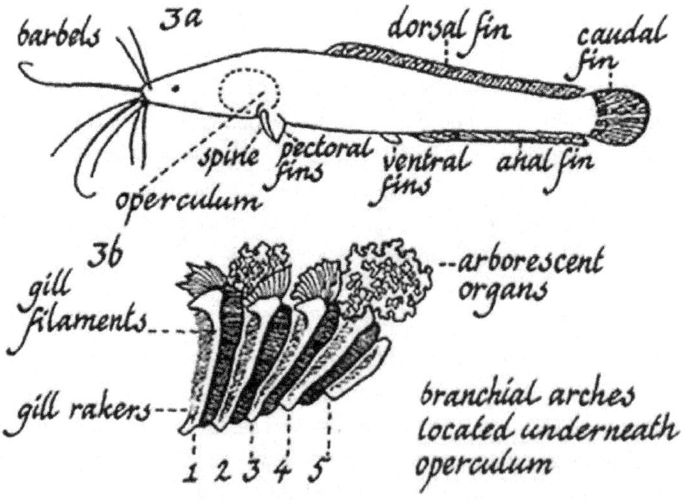

Figures 2 and 3. Drawings of fins, barbels, gills and the arborescent organ of the African sharptooth catfish (source: Practical manual for the culture of the African catfish by W.J.A.R. Viveen et al.).

Gills and arborescent organs

The distribution of the gills and the arborescent organs on the five branchial arches of the African sharptooth catfish, is shown in Figure 3b. They can be observed by cutting away the operculum. For the purpose of respiration, water is taken into the mouth, passes over the gills for gaseous exchange and is then expelled through the opercular opening. Especially when the water's dissolved oxygen content is depleted or if the fish are out of the water, air is periodically gulped in through the mouth. Gaseous exchange takes place via the arborescent organs in air chambers above the gills. The air is also expelled through the opercular openings. Due to its atmospheric respiration, the African sharptooth catfish is capable of surviving in mud during the dry season. It is even able to survive out of the water for some hours depending on the humidity of the environment.

Because catfish can tolerate low oxygen levels in water, they are very suitable for fish culture.

The gill rakers at the back of the branchial arches, serve as filters for feeding on small vegetable matter and invertebrates (Figure 3b).

Anus and urogenital system

In both sexes of catfish, the urogenital opening is situated at a papilla just behind the anus. The sex difference is visible when the fish reach 200 g. The papilla of the male is long and ends in a tip. The female papilla is circular (Figure 4).

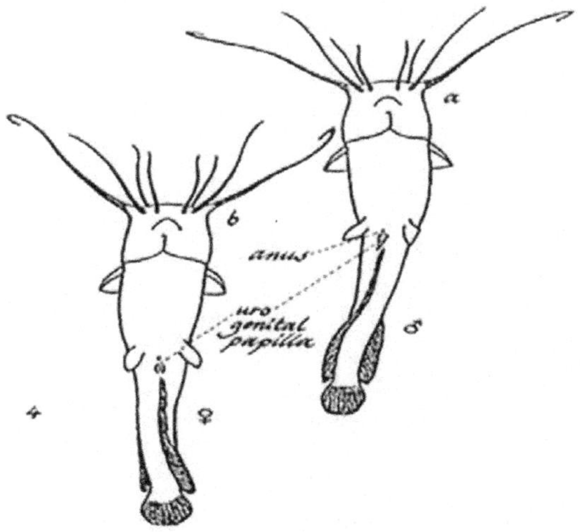

Figure 4. Sexual differentiation between male and female of the African sharptooth catfish (source: Practical manual for the culture of the African catfish by W.J.A.R. Viveen et al.).

2.3 FISH ARE COLD-BLOODED ANIMALS

A cold-blooded animal or ectotherm, like all fish are, does not have an internal mechanism to regulate body temperature. Instead, it relies on solar energy captured by the environment. Catfish will always have a body temperature which is equal to the ambient water temperature. The internal activity of the fish, called its metabolism, increases with increasing water temperature, but it has an optimal range. For the African sharptooth catfish the optimal body temperature varies between 25 and 30 °C.
So, cold-blooded means that a fish is not internally generating energy to regulate its body temperature. Therefore, it requires far less energy than warm-blooded animals or endotherms. Warm-blooded animals need frequent meals on a daily basis to keep the body temperature constant, whereas cold-blooded animals can skip meals for days and even for weeks without any negative health impact.

At our farm, we always take into consideration the temperature of the water and the activity of the fish.

3

THE AVAILABLE SOURCES OF WATER

1. INTRODUCTION
2. WATER FROM A RIVER OR STREAM
3. WATER FROM A DAM OR LAKE
4. WATER FROM AN AQUIVER
5. WATER FROM A WELL
6. CITY COUNCIL WATER

3.1 INTRODUCTION

Catfish farming has to be carried out in fresh water. The salt content (sodium chloride) should not exceed 6 g/l. Unfortunately, most of the water available in the world is salt water. The proportions are as follows: 97.5% of the water reserves in the world consists of salt water (oceans), 1.75% in the form of ice and snow on glaciers at the North and South Poles, 0.5% is underground water and only 0.25% is fresh water from rivers, swamps and lakes.
Africa is the continent with the lowest availability of fresh water; actually, many places in Africa have to deal with a shortage of water. This makes fish farming economically unfeasible. In Nigeria however, we have places with abundant water from rivers, streams and dams, especially in the south and west. In the north we face shortages of water and rivers dry up during the dry season. Underground water is available almost everywhere in Nigeria, but the yield of boreholes and the water quality vary a lot from area to area.
Let us evaluate each source of water.

3.2 WATER FROM A RIVER OR STREAM

Rivers and streams in Nigeria which have water flowing all year round, are very suitable sources of water for fish farming. To find out if the water quality is suitable for fish farming, the first thing to discover is whether the river has an active fishing community. Get your information concerning the fish population and catches from the locals, find out the lowest and highest water level in the river and whether there are any activities which have a negative impact on the water quality and quantity.

Unfortunately, rivers and streams in Nigeria are also used for waste disposal, solid as well as liquid. Another activity which has a really bad impact on the river, is mining. Stones are crushed on the riverside and the river water is used for washing. Chemicals are used to isolate the precious minerals from the sand. As a result, the water becomes extremely turbid and when the flow of water in the river decreases during the dry season, it becomes a poor source of water for fish farming.
Another problem with river water arises with the first rains. When they start, the river water gets polluted with runoff from farmland. The water not only becomes turbid but also gets polluted with insecticides and herbicides. The effect is an increasing rate of catfish mortality, because this pollution weakens their immune system. As a result, bacterial infections are prone to happen. During the peak of the rainy season, the runoff is less poisonous to the fish and consequently the mortality rate reduces. The best time of the year for fish farming in for instance the Abuja region is the peak of the dry season, because the water is clear and less polluted, but this can be different for other regions in Nigeria.

Most rivers in Nigeria become very turbid with every rainfall. The volume of water increases rapidly and decreases rapidly. It's a sign of erosion. The land does not hold any water; all rainwater goes straight into the river via gullies and man-made gutters. Top soil and laterite are washed away into the river, making the soil less suitable for farming. Reforestation is the only way to stop this erosion. Rivers in dense forests don't have this problem, therefore the water turbidity is not affected so much by rain in such areas.

If river water is used as a source of water, there should be provision for an 'intake'. 'This is a place where you are able to pump the water out of the river into a reservoir. The intake should be chosen carefully and should definitely be located above the highest level to which the water can raise during extreme rainfall. The intake should also be installed at a location on the riverbank where the river is deep and the water is close, even during the dry season. This way a very long suction hose is avoided. If a permanent structure is built for the intake, look for locations which have rocks on the bottom of the river, so that the concrete structure built on top of this, has a good and solid foundation. If the farm is small, you can choose a more flexible option, such as a small petrol water pump with a suction hose. For your own convenience, the pump can be placed on the riverbank every day and taken away every night. The water from the river can be pumped directly into the fish pond, but the preferable option is to create an overhead reservoir/tank for collection of the river water. From this reservoir the water can flow to the fish ponds by gravity (Figure 5).

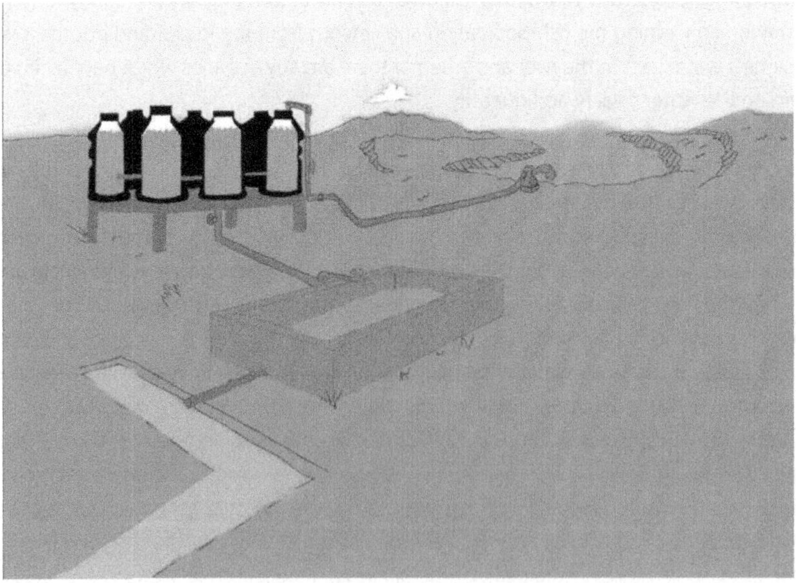

Figure 5. Water, which is pumped from a river or stream to the reservoir, flows to the pond by gravity.

Pumping the water from the river into the reservoir or straight to the fish ponds, can be done either by electrical, petrol or diesel pumps. The choice depends on the availability of electrical power. If there is no electrical power available, a generator has to produce the power for the electrical pump. In this case it is preferable to power the water pump directly with petrol or diesel (Photo 1). Diesel pumps are preferable to petrol pumps, but when the size of the farm is small, you will most likely go for a petrol water pump.

Photo 1. Petrol pump and diesel pump.

The suction hose has to be fixed very tightly in order not to suck any air. A non-return valve should always be placed at the end of the suction hose. The suction hose should be flexible so that it can be taken out any time the current of the river gets too fast and furious. The end of the hose consists of a perforated suction basket which stops any big particles from entering the pump (Photo 2).

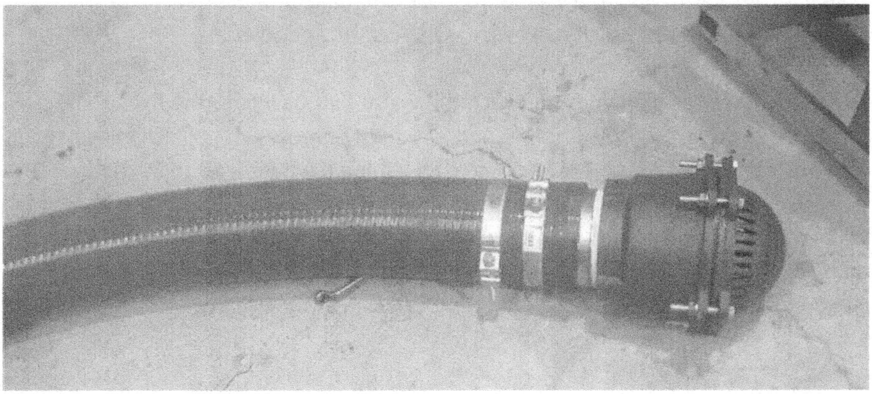

Photo 2. Suction hose with a non-return valve and grid.

Most pumps must be primed before they can operate. This means that the suction hose is filled with water before it can suck by itself. Some diesel pumps are self-priming though, and can be started immediately.

3.3 WATER FROM A DAM OR LAKE

Nigeria has many natural lakes and dams (a manmade lake). The most famous natural lake is Lake Chad. It is a shallow lake located in the north-eastern part of Nigeria and also extends to Cameroon and Niger. The biggest dam is Kainji Dam located in Niger State. The source of Kainji Dam is the Niger River.

Dams are built for many purposes:

- Electricity generated by water turbines
- Irrigation purposes
- Controlling the flow of the river
- Fishing and fish farming
- Recreational purposes
- Water supply for drinking water stations

The water of a dam or lake has some advantages over the water from a river. The quality of the water does not fluctuate as much as in a river, because the dam or lake acts as a very big water buffer. Any deposits from heavy rainfall settle down or are neutralised in the lake or dam. In most lakes and dams the water levels are more stable than in rivers. Furthermore, it's easier to make an intake at a dam than at a riverbank. Dams are usually deep which makes it easy to create a good intake.

Most dams have a pipeline, from which the water falls back into the river, to be used for drinking water stations, turbines and irrigation purposes. This pipeline can supply water by gravity because the level of the land behind the dam is 10 to 50 m lower. This pipeline can also be used for fish farms and saves a lot of money, because pumping the water into the ponds is not necessary. Also, a reservoir is not needed in this case; the dam acts as a reservoir. This situation is the best and easiest way to start a fish farm (Figure 6).

Figure 6. A gravity fed farm.

Not many farms use the option to take water from the pipeline behind the dam. This is because these structures, including the land behind the dam, belong to the federal or local government (water board) and they are reluctant to allow their use for commercial purposes. It really is a missed opportunity, because it could be a very good thing for community development creating fish farm areas for many individual small-scale farmers using the same pipeline and enabling them to enjoy a facility with the lowest possible running costs.

3.4 WATER FROM AN AQUIFER

Underground water can be tapped by drilling a borehole. Depending on the location, depth and size of the borehole, the yield of water varies from 1 m^3/hour to as much as 50 m^3/hour. In order to drill a borehole, first a survey is required to find out the most suitable position for drilling. Underground water is nothing less than collected rainwater which flows via the permeable soil into the aquifers. Aquifers are underground layers of water-bearing rock. This process has been going on for centuries. Some of our groundwater can be as old as 1000 years. In recent years, due to urbanisation, a rapid increase in the number of boreholes is depleting the underground (borehole) water rapidly. Fish farmers can use borehole or underground water, but only for the first stages of fish farming, from the hatchery to the post-juvenile stage (see Appendix 1 for the catfish growth stages), in order not to deplete the reserves of underground water for future generations.

What are the constraints of boreholes in general?

- A company drilling boreholes will never guarantee the yield and the quality of the water, but will ask you to pay at least 70% upfront.
- In most areas of Nigeria boreholes give only a moderate yield of less than 5 m^3/hour. With this yield a small grow-out farm is possible but it can never serve a medium or large fish farm.
- Pumping with a borehole pump takes quite a lot of electrical power. In addition, borehole pumps can break down or the borehole tends to dry up.
- A fish farm needs at least two boreholes.
- The yield of the borehole does not depend on the size of the pump but on the recovery rate of your borehole when pumping. For example, when the water level in your borehole is at 10 m depth when you start pumping, and the level drops by 10 to 20 m depth, while a small pump of only 1.5 hp is used, it means that the recovery rate of the borehole is very poor. The water level of a good borehole hardly drops when water is pumped out. The yield is very high and the aquifer has a lot of water which can be easily taken out.
- Some boreholes contain high levels of iron or bacteria which make them unsuitable for a hatchery and/or even a grow-out.
- Borehole water has very low levels of oxygen and high levels of carbon dioxide. Before the water enters the fish farm, it should be exposed to air in order to become rich in oxygen and low in carbon dioxide.

What are the advantages of boreholes?

Many boreholes supply water of drinking water quality. That makes them suitable for hatcheries. The water quality of a borehole is very stable as long as the borehole is drilled well and the soil around the borehole pipe is sealed well.

Because river water is unsuitable to use for hatcheries in almost all cases, the only option for a hatchery is a borehole or water from the water board. The use of drinking water is only possible as long as you allow the chlorine to escape within 24 to 48 hours.

How to keep boreholes clean?

The borehole drilled on your land should always be located far away from waste disposal areas and soakaways. Keep the environment where your borehole is situated, as clean as possible. Test the quality of the borehole water every two years for chemical content and at least once a year for coliforms, which are bacteria of faecal origin which indicates the presence of other pathogenic bacteria. Also clean your water reservoir at least four times a year to avoid growth of unwanted bacteria and algae and remove silt deposits from the bottom of the reservoir.

3.5 WATER FORM A WELL

A well is an open pit, dug into the earth, deep enough to reach the groundwater level. The water can be pumped or taken out with a basket and rope. The yield of a well is limited and only considered useful for domestic use, not for fish farming.

3.6 CITY COUNCIL WATER

City council water cannot really be a reliable source of water for fish farming. Also, the water is chlorinated and toxic for the fish if it's used directly from the tap. The chlorine causes the slime on the skin to dissolve which will definitely kill the fish. Chlorine however is a gas. When the water is exposed to air and light, the chlorine will evaporate within one day after which you can use the water for fish farming.

4

FISH CULTURE SYSTEMS

1. FLOW-THROUGH SYSTEMS
2. RECIRCULATION SYSTEMS
3. CAGE SYSTEMS
4. EXTENSIVE POND CULTURES
5. BACKYARD FARMING

4.1 FLOW-THROUGH SYSTEMS

The majority of fish farms run a system called flow-through. Water comes in and goes out, carrying the fish waste out of the pond (Figure 7). The water is not meant to provide oxygen, since catfish, apart from the early stages, take in the majority of their oxygen from the air.

Figure 7. Schematic drawing of a flow-through system.

Use of water

Fish farming is all about water management. In this context it is very strange that most farmers don't know how much water they have access to in a day or a year. The quantity of water used in a day, is directly related to the fish production. Almost every fish farmer thinks otherwise. They believe it's the number of fish ponds that determines the production capacity. However, filling the ponds is just the beginning. The water in the pond has to be flushed constantly in order to keep the water quality acceptable so that the fish can enjoy the environment, eat and grow well.

If the relationship between the size of the pond and the number of fish is not relevant, how do I calculate the number of fish I can stock in my farm?

The stocking density in a pond is related to the flow of water which passes through the pond. In other words, the more water you are able to flush out of the pond, the more fish you can stock in that same pond. To be more precise, the flushing rate is directly related to the amount of feed that is given. This is related to the number and size of the fish in the pond; the more fish you stock, the more feed you need.

The feed is consumed by the growing fish, but this process also causes pollution in the form of faeces, urine, ammonia and carbon dioxide. All these accumulating waste products will cause serious toxic problems for the fish. Therefore, they have to be removed. This is only possible by flushing the water periodically or constantly out of the pond.

The relationship between the feed and the flushing rate is as follows: for every 1 kg of feed given to the fish, you need to flush 1.4 m^3 of water.

How do we know this equation?

The equation for feed and flushing is based on practical experience, but also backed up by measuring the ammonium level which should not exceed 20 mg/l. Ammonium is the first limiting factor to cause problems for the catfish. Because without flushing the ammonium level increases every day, flushing has to take place preferably on a daily basis.

Parameters that will change the equation of 1 kg of feed to 1.4 m^3 of water:

- Overfeeding: the feed is not being eaten completely by the fish, polluting the water directly.
- The feed is of poor quality and the digestibility of the feed is low: the fish will produce more waste per kg of feed consumed.
- The feed is not well processed: it falls apart in the water very quickly.

The equation is based on good quality feed and optimum feeding.

Example 1. How to calculate the flush rate for fish of 0.5 kg?
Suppose we have 1000 fish, having an average weight of 0.5 kg each. If we feed 10 kg (2% of biomass) of feed in a day, how many litres of water do we need to flush? Recall that for every 1 kg of feed, we need to flush about 1.4 m^3 of water.
Hence, we need to have a flushing regime of

$$= 10 \text{ kg /day} \times 1.4 \text{ m}^3\text{/kg}$$
$$= 14 \text{ m}^3\text{/day}$$

The fish can be stocked in 10 m^3 of water (100 pieces/m^3) or 20 m^3 of water (50 pieces/m^3) or 100 m^3 of water (10 pieces/m^3). All options will work because catfish are very flexible when it comes to stocking ratio, but the flushing has to be 14 m^3/day for all pond sizes. The same fish will grow, but when they become bigger, they subsequently need more feed, which means that the flushing rate should be adjusted.

Example 2. How to calculate the flush rate for fish of 1 kg?
Suppose we have 1000 fish, having an average weight of 1 kg and they are fed 15 kg per day (1.5% of the biomass). What would be the flushing regime (rate)?
Recall that for every 1 kg of feed, we need to flush about 1.4 m^3 of waste water.
Hence, for feeding 15 kg in a day, we would need to flush

$$= 15 \text{ kg/day} \times 1.4 \text{ m}^3\text{/kg}$$
$$= 21 \text{ m}^3\text{/day}$$

As you can see, when the fish get bigger, the flushing rate increases drastically.

Many farmers start with stocking juveniles. At that time the flushing rate is low because the feed consists of just a few kilos a day. The problem starts after some months when the fishes' need for feed has really increased. The flushing rate has to increase drastically. When the flushing rate does not keep up with the increasing feeding levels, the water turns from green to black. This is a sign of anaerobic conditions and the water starts to smell. The fish will start to hang at the surface and refuse to eat as before. This condition happens to many farmers and it's all due to poor water management.

Efficiency of flushing water

The efficiency of flushing a concrete tank, pond or any other tank varies from the way you do it. If the tank is full and flushing is done by causing the water to overflow, your efficiency is lower than when you drain part of the water and replace it with fresh water. The efficiency of replacing water is two times higher than the first option (Figure 8).

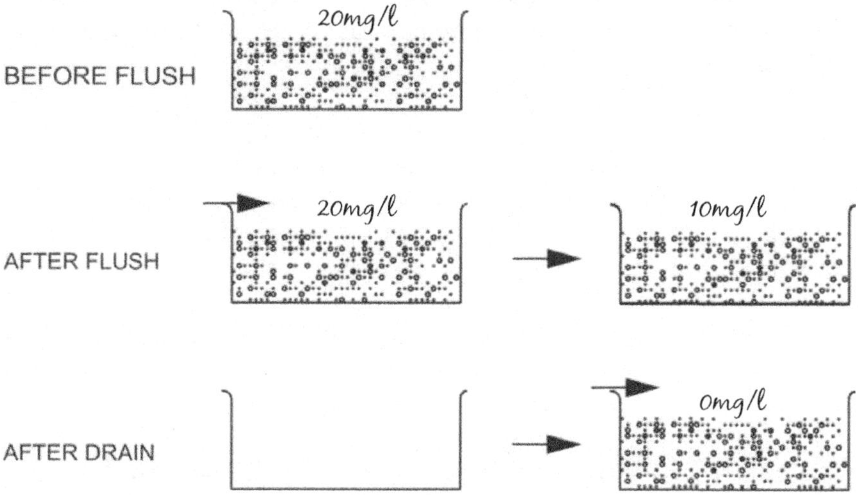

Figure 8. Flushing or draining the water from a pond or tank.

Replacing the water has one disadvantage!

Replacing the water from a pond completely or almost completely, is very efficient in terms of improving the water quality straight away, but it will stress the fish, which can lead to increased mortality and higher feed conversion rates. This practice is done by many farmers; they allow the water to get completely bad and smelly, resulting in fish hanging at the surface not eating well and mortality increasing, before they change the water completely.

In my opinion, the best way of flushing the water from a tank is by lowering the water level 20 to 30% every morning and allowing fresh water to come in and overflow from the tank for the rest of the day.

Another malpractice, which is carried out by many farmers, is to stock all the tanks in a short period of time. This will lead to an extremely high demand for water when the fish all reach their market size at the same time. Consequently, this leads to hardly any need for water after all the fish have been sold.

Remember the flushing rate is directly related to the feed given!

Example 3. Why does spreading the production and the sales over the year have more advantages than selling twice a year?
Suppose that the production is 10,000 kg catfish a year.
Then the total water consumption for that year = 10,000 x 1.4 m^3 = 14.000 m^3

Case 1. If the production is spread over 12 months (approximately 830 kg/month) the consumption of water is equal to the maximum 14,000 m^3/365 days = 38 m^3/day or 1.6 m^3/hour.

Case 2. If the production is done in two cycles of 5,000 kg, we will need the same quantity of water in a year as in case 1, but we have a peak consumption at the selling point, which is around 105 m^3/day or 4.4 m^3/hour: 5000 kg of fish before sales get 1.5% of bodyweight: 75 kg. The ratio is 1.4 m^3/kg of feed. So the water consumption is 75 kg x 1.4 m^3/kg of feed = 105 m^3.

The consumption of water at this point is three times higher than if we spread the production on a monthly basis. Hence, spreading the production over the year has many more advantages, which will be discussed in other chapters.

Example 4. How to achieve a constant supply of water without a permanent power supply or running water source?
Suppose a farm has a capacity of 10 tons (10,000 kg) production of catfish a year. The question is how big should the reservoir be, if you want a back-up for 24 hours?
First, we have to calculate the water usage (consumption) in the fish farm.
Secondly, we calculate the flow rate of the pump which pumps the water to the reservoir.
A fish farm ideally needs a constant supply of water. The only way to achieve this is by a water reservoir.

We know that 1 kg of production needs 1.25 kg of feed (feed conversion rate is 1.25). This will be further explained in chapter 7, which deals with feed.
Flushing of water: 1.4 m^3 per kg of feed fed.
Total feed per year required for this production is 10,000 x 1.25 kg of feed = 12,500 kg.
Total water required according to the former equation: 12,500 x 1.4 m^3 = 17,500 m^3 in a year.

Total water required per day is 17,500 m^3/365 = 48 m^3 or 2 m^3/hour. Suppose the pump has a flow rate of 6 m^3/hour, we need 8 hours of pumping to supply the daily need for water: 8 hours x 6 m^3/hour = 48 m^3.

During 8 hours water can be pumped to the reservoir and to the fish ponds at the same time. 16 m^3 goes to the fish ponds, and the remainder 32 m^3 flows into the reservoir. The storage reservoir should be exactly 32 m^3, but with some margin it could have the size of 40 m^3 (4 tanks of 10 m^3).

The logic behind this equation is that, when the supply to the reservoir is high (flow rate), the number of hours to supply the daily requirement is limited, which means that the storage capacity has to be bigger. On the other hand, if the flow rate is 2 m^3/hour, you don't need a reservoir, because the supply is equal to the consumption. When the pumps stop, there would be no water in the reservoir at all.

Many farmers think that a big reservoir enables you to start a big farm, but that's a wrong assumption. A reservoir is only useful if the supply exceeds the demand. If the demand is higher than the supply, it's a simple fact that your farm is not well balanced. You have to reduce the number of fish ponds, reduce the stocking of fish or increase the water supply.

Pond design and water management

A well-constructed pond built of either concrete, plastic lining, earth or otherwise, should have a well-designed drain which is easy to manage (Figure 9).

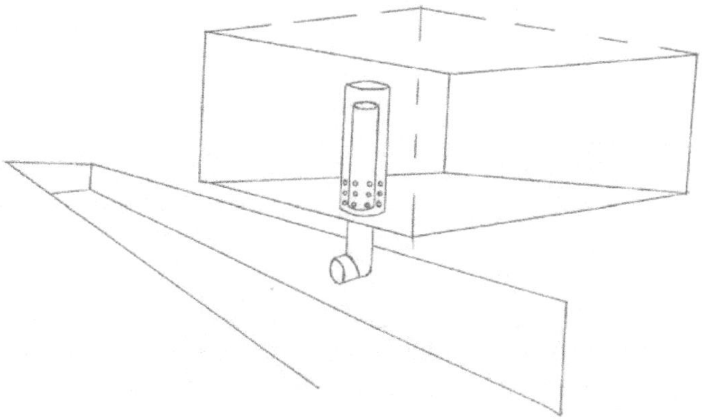

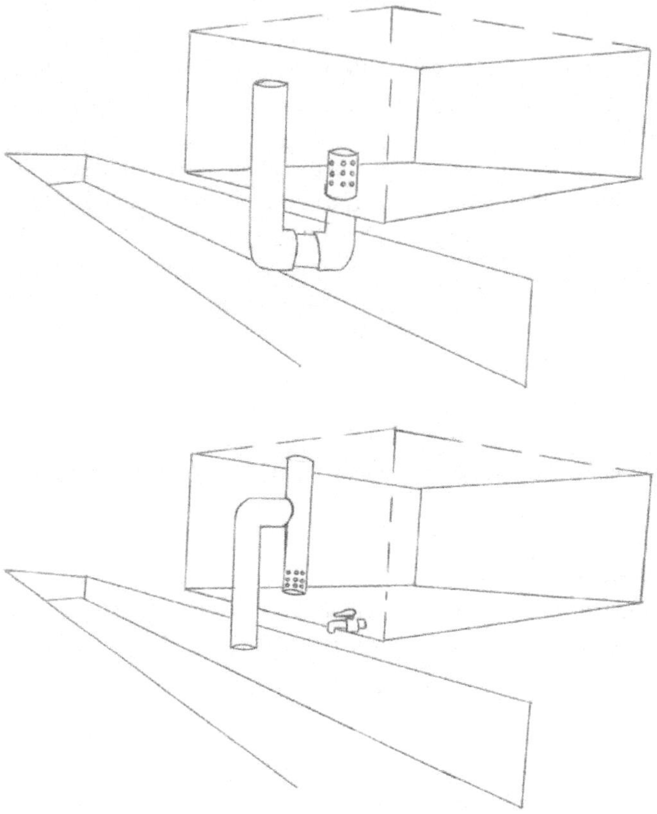

Figure 9. Three ways to drain a pond.

The inlet to the pond should be via a pipeline which is connected to the reservoir, directed to the ponds and controlled by a valve.

In many farms the inlet and outlet (the drain) of the ponds are too small. As a result, the water will not be drained and filled quickly. The main inlet and outlet pipes for ponds of 20 m^3 or more, should have a diameter of 4 inches or 110 mm. The drainage should also have a diameter of 4 inches or 110 mm.

Ponds can be built completely dug into the ground, half way dug into the ground or completely on top of the ground. All depends on the drainage. The drainage should go by gravity. Pumping out the water to flush the tank, is not an option and should be avoided by all means.

One exception to this matter: ponds constructed in marshlands, can only be drained by pumping out the water. Fresh water automatically comes in from underground and an intake is not required. During the rainy season this process is very quickly, in the dry season recovering water takes time. These ponds can only run extensively, but it's possible and feasible. Emptying the ponds during the rainy season is difficult though, because fish get left behind in the mud and interfere with the next stocking.

The waste water of the ponds can be directed to a lower collection reservoir (settling tank). The overflow of this collection reservoir can go into a canal, river or a gutter, which leads to a river or stream. Never use a soakaway to collect water, because it will clog up in a short period of time and then the water cannot go anywhere. Drainage water should not be going to other people's land to avoid clashes with neighbours. Waste water is useful for irrigation, but this is only applicable during the dry season when the fish farm is surrounded with enough land.

Ground water

If you build ponds in the ground and the ground water level can reach the bottom of the pond, it can cause severe damage to the pond. When the pond is empty and the ground water level reaches above the bottom of the pond, the pond can be lifted and it can break. In the case of plastic lined ponds, the pond liner will be lifted and it's like you see a hippo inside the pond. For the earthen pond a higher level of ground water is not a problem, but you cannot drain and dry the pond because the ground water will keep on coming. If you suspect the ground water level to be very high, always keep the plastic ponds full of water, even if there are no fish inside.

Roofing the ponds; concrete ponds in particular

Many farmers like to roof concrete ponds. I believe this just adds to the investment costs without any benefits. Farmers believe that the direct sunlight will increase the water temperature to a level which will harm the fish. In concrete tanks the water temperature does not increase very rapidly though. When you flush on a daily basis, which is necessary, the incoming fresh water will reduce the water temperature to levels below 30 °C. An additional benefit of having no roof over the pond, is the growth of green algae. Green algae are really beneficial to the catfish; they like this green environment and are less susceptible to bacterial diseases. Catfish grow even more quickly in green water compared to clear water. Additionally, the ammonium levels in green water will also reduce, because algae take in the ammonium. As a result, the flushing rate can even be lower than in fish ponds with a roof. With a roof the algal production is inhibited because sunlight cannot penetrate the water directly.

There is however one exception to not roofing the fish tanks. That is when you use polyethylene (plastic) tanks. These tanks are mostly used to grow juveniles to post-juveniles (see Appendix 1 for the growth stages of catfish). If situated on a floor in the sun, the water inside the plastic tank can become over 35 °C which is too high for the fish. When plastic tanks are used, it is advisable to roof them to prevent exposure to direct sunlight. Covering the top of the tank with banana leaves is a cheaper and also an effective option to avoid high water temperatures.

Water level in a pond

The water level in ponds should not be less than 100 cm. Many farmers however have water levels of 30 to 40 cm or even less. This causes unnecessary stress to the catfish because the exposure to sunlight is high; in addition, the fish cannot hide and the temperature fluctuations in ponds with a low water level are also large.

Pipes

The pipes for drainage and overflow used in concrete ponds, should be of premium quality. Use only high-pressure pipes of 10 bar. The bottom drain should end with a socket, so that you can put a standpipe inside it. To improve the connection between the PVC and the concrete, you should sandpaper the sockets and the pieces of pipe to be put in the concrete. After sandpapering, put PVC-glue on the sockets and attach fine sand. This way, the connection between the pipe and the concrete will be strong, and without any leakage. If your underground piping leaks at the connections, it will be very hard to fix. Besides, the leakage will eventually weaken the ground underneath and destroy your pond. So, don't minimise costs when it comes to pipes and fittings for your concrete system. it's a long-term investment.

The size of the pipes needed is related to the stages of the growth cycle of the fish. In grow-out ponds the size should be 110 mm diameter or more, for pre- and on-growing 75 mm, and for juvenile production 50 mm. If the pipes are too small, they will not allow you to flush and supply on time and they can easily become clogged up.

Construction of a reservoir

Reservoirs should be built at least two metres higher than the upper level of the fish pond. This allows the water to flow by gravity from the reservoir to the fish pond.

A reservoir can be built in many ways. Depending on the size, you can choose plastic tanks on a scaffolding (the biggest are 10 m^3 each) or a block wall pond on a higher level of the terrain (Photo 3). Using stone pitching or plastic lining to build the reservoir is also possible (Photo 4). If you have a big property, it is important to do a good survey on your land to ensure that you build the reservoir on the right spot, which should be at least two metres higher than the upper level of the fish pond, and as central as possible to avoid an excessive number of pipes.

The piping from the reservoir to the fish pond has to be done with good quality PVC or polyethylene pipe. The diameter of the supply pipe from the reservoir to the fish pond varies with the stages of the growth cycle: for hatcheries a minimum of 50 mm is required, for grow-out the diameter should be at least 110 mm. Depending on the farm size and the distance to be covered by the piping, the size of the pipes will increase to 160 mm, 200 mm or even 250 mm. A qualified engineer has to provide you with the dimensions of the piping system, taking into consideration the flow rates, levels and distances.

The pipes should be preferably installed underground and the surveyor should map the pipes on a drawing, so that they can always be located in order to avoid damaging them when further construction takes place.

The hatchery water reservoir needs special attention; it has to be cleaned and drained periodically, advisably every three months, in order to avoid any accumulation of debris. If the tanks are dirty, disinfection with chlorine is even required. After washing and rinsing, you are good to go again.

If borehole water has a high concentration of magnesium and/or calcium, deposits of these metals easily form on the walls and inside the pipes. These deposits can harbour bacteria which will make the water unsuitable for hatching and even for small fry (see Appendix 1 for the catfish growth stages). When the water is very soft, you will notice that all your tanks and equipment are easy to clean and look brand new afterwards. This is the best water for a hatchery.

Living in Nigeria on an estate, we have to be self-supporting for water, power and security. This is in itself a daily job, because all these factors need control and maintenance. At one point our entire family was constantly sick and we were wondering what was happening. We all went to a laboratory and found out that we were infected with salmonella. The origin of the infection was our water tank. The water tank was dirty and dead lizards had made it even worse. After cleaning and chlorinating the water tank, we quickly recovered. Since then, cleaning the water tank has become part of our regular maintenance.

Photo 3. An elevated plastic tank reservoir.

Photo 4. A large reservoir with plastic lining.

4.2 RECIRCULATION SYSTEMS

In a recirculation system (Figure 10) the water is not used only once, but is also recycled via a water purification process and flows back into the pond. Recirculation is not 100% but can be as high as 95% depending on the quality of the water purification system and the management applied. This recycling process means that the water consumption per kg of production is drastically reduced. Instead of 1.4 m^3 per kg of feed, only 0.2 m^3 per kg of feed is required.

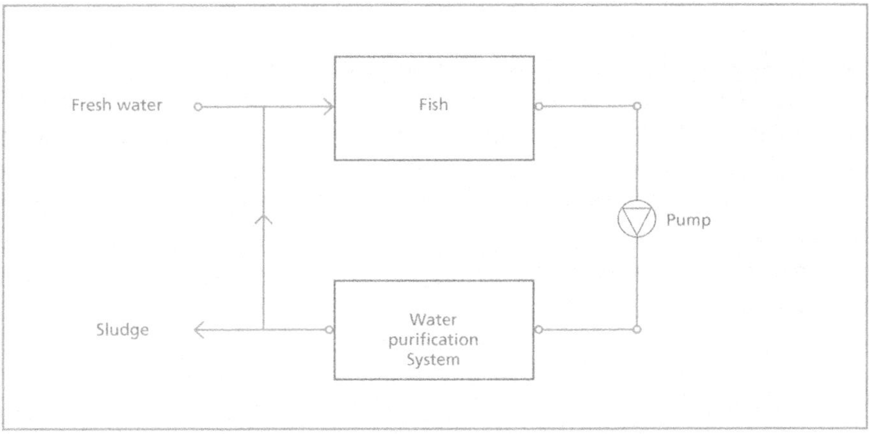

Figure 10. Schematic drawing of a recirculation system.

Recirculation of the water however, comes at a price. The system can only operate with 24/7 power supply, because pumps are running 24/7 in order to push the water through the purification system back to the fish ponds.

It is a high-tech system and it can only be operated by farmers who have the engineering skills and practical experience to work with a system like this.

In the early 2000s (2000 - 2005), many farmers wanted to grow fish in a recirculation system. Their reasons were:

- Farming in a recirculation system (Photo 5 and 6) can be done very intensively and doesn't need much space. So, farming can be done even in town or in an industrial area. Most investors and farmers don't live on their farm, but want to live close to the farm for routine checks, control etcetera.
- Farming in a recirculation system close to town or in town, offers the opportunity to have customers close by.
- Since it's a very intensive system, you can secure it and control theft much better than with an extensive pond farm system, which was used mostly at that time.
- Because the majority of the water is recycled, the farm can run with one or two good boreholes, making the farm independent of other sources, such as a river or a dam.

Although many farmers started very enthusiastically with recirculation systems, evaluation of the systems in 2018 shows that over 90% of them have left the recirculation system and closed down the farm or run the system as a flow-through system.

Why are recirculation systems not popular anymore?

Nigeria has not improved when it comes to a stable power supply. Running a generator any time power failure occurs, is very challenging. Therefore, farms mostly run the recirculation system with power on and off, but then it is not a recirculation system anymore; you can compare it with half pregnancy – either you are pregnant or you aren't! Half pregnancy doesn't exist!

Furthermore, the running costs are higher than expected and the system needs dedicated staff to keep it working well. Even theft of a recirculation system is not excluded. The reason is simple: most of the theft is done by insiders.

Theft: The fact that a farm is compact and controllable, is not an advantage if staff are the ones stealing.

In areas where the nights are relatively cold or during harmattan, the biofiltration system cools down the water drastically. The temperature of the water can drop more than five degrees overnight. This is not good for the health and growth of the catfish. This is not really a problem in areas such as Lagos and Port Harcourt, but it will be a big problem in a place like Abuja.

Farmers who run a recirculation system, eventually have to compete with other farmers and if their cost price is higher, it will be difficult to make a profit when the market is competitive.

The water purification system for the recirculation system is not available in Nigeria, so it has to be imported. Because the naira has become very weak against the euro and the dollar, these items became very expensive.

There is however one exception when it comes to running a recirculation system for fish. That is when it is used for hatchery operations. Hatchery recirculation systems are very effective and competitive.

Photo 5. A recirculation system for African catfish grow-out in the Netherlands.

Photo 6. A recirculation system for an African catfish hatchery.

The operation of a recirculation system and its principles

Because a recirculation system works well for hatcheries, I will explain more about the operation of recirculation systems and their principles.

In order to design a recirculation system for fish, we have to know the type of waste and the quantity of waste a fish produces. The waste of the fish starts to build up from feeding onwards.

The following equation can be used:

Growth (of the fish) = input (feed) – output (waste)
Feed--------feed intake and feed loss
Feed intake----------digestible fraction and faeces
Digestible fraction---------metabolic fraction and ammonium (urine)
Metabolic fraction--------growth and respiration losses (carbon dioxide and heat)

The fish waste is: uneaten feed, faeces, urine in the form of ammonium, carbon dioxide and heat.

1 kg of feed will result in the following waste (based on dry matter):

- 135 g of faeces
- 43 g of ammonium-nitrogen (most of it in a soluble form)
- 9 g of phosphate (most of it in the faeces)

The waste is valuable for other agric purposes but the problem is, it is packed in a lot of water! Chapter 13 tells you more about using waste water for crops.

Which processes can be used to eliminate the fish waste?

The fish waste accumulates in the water if it is not eliminated and eventually the water becomes toxic for the fish. In the flow-through system we simply flush it out before the water becomes toxic, but in a recirculation system we apply a purification system and re-use the water.

3 processes for dealing with fish waste:

1. Sedimentation of the waste water in order to separate the suspended solids (faeces) from the water and drain this fraction out of the system. If the solids stay in the water, they will cause anaerobic processes which produce toxic gases, for example H_2S. Excessive bacterial growth and solids in suspension also have a negative impact on the gills of the fish. Filtration is another way to reduce the suspended solids.

2. Nitrification of ammonium by bacterial processes in order to eliminate ammonium and change it into nitrate, which is not toxic for the fish. High ammonium levels in combination with a high pH can cause severe damage to the slime and skin of the fish. If the slime and skin of a fish are affected, it will certainly die.
3. Removal of the carbon dioxide from the water to the air by degassing. High carbon dioxide levels are toxic for fish. They make them sluggish. The fish can suffocate and eventually die.

Systems to facilitate the waste processes

Sedimentation tank

Sedimentation is a relatively easy process. The flow rate in the sedimentation tank (Photo 7) is slow and allows the particles which are heavier than water, to settle before flowing out. As a result, the particles will settle onto the bottom. When the tank includes sedimentation filter materials, it increases the settling rate and trapping of solid particles, which results in overflow water without any particles from the suspended solids. The next step is to regularly flush and rinse the sedimentation tank, otherwise the solids start to dissolve and have a negative impact on the nitrification process. By flushing the sedimentation tank, you also refresh the water. As a result, the salt levels will not keep on increasing but stay in balance. A good sedimentation tank can reduce the load of fish waste by 50%. A sedimentation tank can also include bacterial processes such as denitrification. This occurs when oxygen levels in the tank are close to zero and the organic load (faeces) is eaten by bacteria which in their turn produce nitrogen gas out of nitrate.

- **Equation** $2NO_3^- + 12H^+ \rightarrow N_2 + 6H_2O$

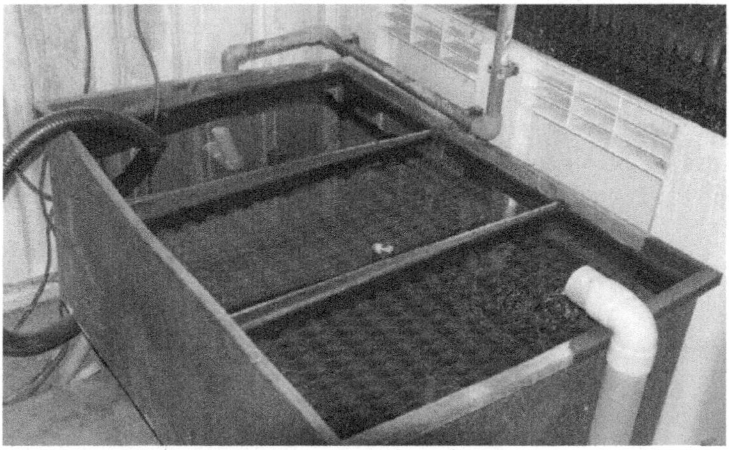

Photo 7. A sedimentation unit for a hatchery.

Filtration unit

Filtration of suspended solids takes place when the mesh size of the filter material is smaller than the particles. The particles are trapped by the filter. The filter has to be backwashed regularly in order to avoid complete clogging, so that even water cannot pass through. Backwashing also involves flushing of some water. A common filtration unit, which is used in almost every swimming pool, is a sand filtration unit. Another useable filtration unit is a drum filter, so called because it looks like a drum. In hatcheries you rarely see filtration units, because almost all hatcheries with recirculation systems, practise the sedimentation process.

Nitrifying unit or biofiltration unit

The nitrifying unit operates by a biological process. With the help of bacteria, ammonium, produced by the fish and excreted via the gills, is converted into nitrite and nitrate respectively (Photo 8).

The equation is as follows:

Ammonium with oxygen is converted into nitrite by bacteria (Nitrosomonas)

- Chemical equation: $2NH_4^+ + 30_2$-------- $4\ H^+ + 2H_20 + 2N0_2^- +$ energy

Nitrite with oxygen is converted into nitrate by bacteria (Nitrobacter)

- Chemical equation: $2NO_2^- + 0_2$---------$2NO_3^- +$ energy

What you need to know about this process is the following:

Converting ammonium in nitrate is a two-step process. The intermediary nitrite is a very toxic metabolite which can kill the fish. Nitrate is not toxic for fish even in concentrations as high as 500 mg/l. But nitrite already kills small catfish at a level of 3 to 4 mg/l. When the biofiltration unit is not well developed, the water will always contain high levels of nitrite which can only be neutralised by flushing the water. The nitrification process makes the water acid. In a situation in which the borehole water has no buffer, it is advisable to add sodium bicarbonate to the water from time to time to keep the pH at 7 (neutral). The bacteria doing this job, need a surface area they can attach to (substrate) and enough oxygen to be able to convert the ammonium into nitrate.

Photo 8. A biofiltration unit.

Photo 9. Different filter media.

What you should know about filter media:

- The material from which the filter media is made (Photo 9), has to be non-toxic for fish.
- The surface area has to be large (200 m^2/m^3), but should be open enough for self-cleaning. Excess bacteria on the media should be washed out with the water that flows through it. For example, a stone bed is a good medium in the beginning, but when bacteria start to grow on it, it will clog completely, which will be very difficult to clean.
- Air and water have to pass through the media easily so that exchange of carbon dioxide and oxygen can take place efficiently.

The biofilters can be set up as a dry filter system; the water is pumped through the filter as a rain shower. Or, the biofilter unit is a submergible filter and the oxygen is supplied by an air pump.

Extra note on nitrite intoxication

Small fish like fry are extremely sensitive to high nitrite levels. The first signs of intoxication are sluggish behaviour and hanging at the surface of the water. Their appetite reduces and after some hours or a day the first mortality is recorded. If not properly addressed, most of the fry die within 48 hours. Concentrations already start to be toxic at 2 to 3 mg/l. When you have a recirculation system for a catfish hatchery, you need to use a test kit to test the nitrite level in the water regularly.

When the catfish are bigger and the nitrite level suddenly rises, the fingerlings or juveniles can die very quickly too. This is accompanied by a very specific sign. The blood of the fish gets brownish. This is because of the reaction between the blood and the nitrite. The haemoglobin in the blood binds to the nitrite which results in the blood turning brown. This also explains the toxicity of nitrite, because it makes the transport of oxygen in the blood almost impossible and the fish will suffocate.

2 remedies are possible:

1. Changing the water with clean borehole water in order to reduce the concentration of nitrite drastically.
2. Adding 3 kg of sodium chloride (industrial salt) per m^3 of water to neutralise the nitrite and reduce its binding capacity to haemoglobin.

You can also add the salt after flushing. Combining the remedies makes them even more effective. Safe levels of nitrite for catfish are below 1.5 mg/l. Check every day and adjust on a daily basis if necessary. Industrial salt should always be available as standard material in a hatchery!

Degassing unit in order to remove carbon dioxide

When the fish metabolise, they produce carbon dioxide. This gas will partly be dissolved in the water which causes a problem at high concentrations. Carbon dioxide tends to make the fish sluggish which reduces their ability to eat. Compare it to a poorly ventilated room which you are in. The only thing you want to do is sleep. The carbon dioxide is easily removed with a properly functioning biofiltration unit. In the biofiltration unit oxygen will diffuse into the water and carbon dioxide will diffuse into the air. When the water has passed this unit, it is in a good condition for the fish and will go straight to the fish tank.
No special unit is needed, because a properly functioning biofiltration unit will automatically take care of degassing the water.

UV-filtration unit (UV = ultraviolet)

The UV-filtration unit reduces the level of pathogenic bacteria in the water. A UV-light provides ultraviolet rays which kill bacteria by changing their DNA (Photo 10). However, UV-light is dangerous for your eyes, so never look at UV-light when it is on! UV-rays only work when the water is transparent, allowing them to pass through, and if the flow rate of the water is in line with the capacity of the UV-light. The filtration unit should be cleaned and its functioning should be checked after every cycle. In hatcheries it can be especially helpful to keep a healthy environment for the catfish fry. The UV-system should be configurated in such a way that, after the water has gone through the sedimentation system, it is free from particles when it passes through the UV-filtration unit.

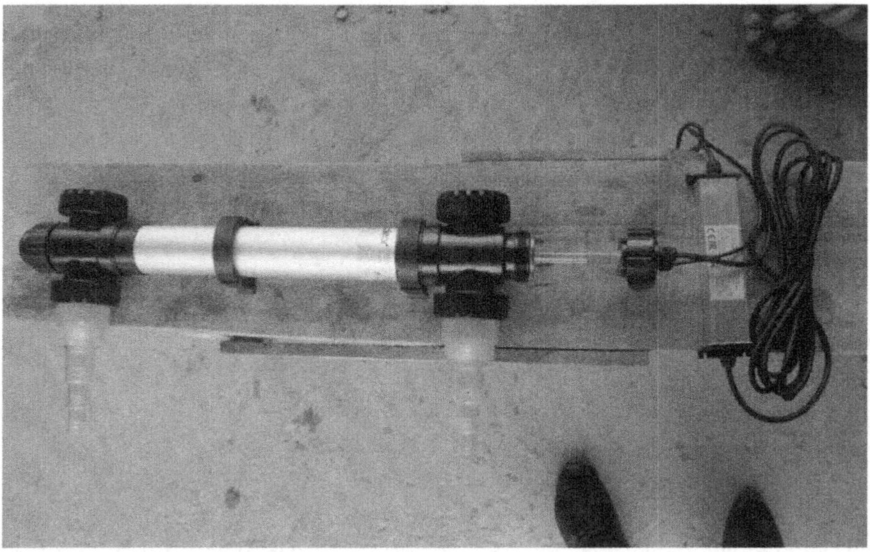

Photo 10. A UV-filtration unit.

Starting up a recirculation system

A brand-new recirculation system has no bacteria inside the filter. The good thing is that bacteria will move in spontaneously the moment you run the system and fish release their waste products. These are food for the bacteria. It takes about six weeks to develop a fully-grown bacterial mass on the filter media. During this period, the levels of ammonium become high in the first two weeks. Thereafter high levels of nitrite levels arise for about three weeks. In this period, you have to flush a lot of water in order not to poison your fish. Also start with small fish or fry so that the load of waste is small and the bacterial mass can grow alongside the fish.
You can also buy nitrifying bacteria in a bottle and add them to the system. This start-up will be faster. Another method is to start up the filter without fish by just adding a little feed or urea into the system and allowing it to run for some weeks. The third method consists of putting a spare filter media in a tank submerged in the water aerated and fed with some fish feed or urea. The filter media, when covered with bacteria, are slimy and can be used in a system to speed up the nitrifying process.

The water in a good recirculation system will become brownish. This is the effect of recirculation of the water combined with some ingredients in the feed, such as blood meal, because they are not completely decomposed by the bacteria. This process is not harmful for the fish; as a matter of fact, catfish like to swim in this tea coloured water. It reduces stress and helps to prevent diseases.

If the water in the recirculation system is foaming, it is a sign of pollution by dissolved proteins in the water. This is not really a problem, but it shows that the water purification system can barely cope with the pollution. If there is excessive bacterial growth on the filter media, it is a sign that the system is overloaded. It can also be a sign of overfeeding. Uneaten feed has to be decomposed by bacteria which are different from the nitrifying bacteria. As a matter of fact, they will occupy all the space and nitrification doesn't take place any more. This process, called mineralisation, consumes a lot of oxygen and produces excess bacterial sludge. This process will be enhanced when the sedimentation filter is overloaded and not cleaned regularly.

Medication and recirculation systems

In a recirculation system you can medicate the fish, but depending on the medicine or chemicals you use, this can affect the performance of the biofilter.
Treatment with chemicals, such as formalin and potassium permanganate, will kill some of the bioflora on the filter media. After treatment, extra flushing is required to keep the levels of ammonium and nitrite low. Normally the bacterial mass will return within a week.

Other medicines, such as antibiotics, usually have no effect at all on the bioflora of the media. They specifically kill the pathogens and not the bacteria in the filter.

Salt, widely used to suppress bacterial growth and parasites, has no effect on the bioflora of the media.

When the fish are given medicines via a bath treatment (see chapter 9 on fish diseases), it's advisable to treat the complete system and not only the culture tank. Apply the chemical in the fish tank, but keep the pump running. By doing it this way, you avoid the fish getting stressed. If a chemical like formalin is used, you have to increase the flushing rate in order to remove the formalin after some hours.

The effectiveness of the medicines usually reduces with time, because they are not stable in a water solution and the filter breaks them down too. Especially oxytetracycline which breaks down rapidly. The water will turn dark brown within twelve hours after applying it. Don't bother to flush it out, since the catfish like this dark tea coloured water.

Maintenance of a recirculation system

A recirculation system requires maintenance. The major maintenance consists of keeping the components clean. When the pump gets dirty, the yield of the pump reduces gradually. It's important to clean the impeller on a monthly basis or every time the system is emptied after a growth cycle. If the system uses UV-light, the quartz glass should also be cleaned after every cycle, and checked if the UV-light still works. Ultraviolet radiation kills pathogens and keeps the water transparent, but if the quartz glass of the UV-light is dirty, the radiation is not optimal and will not do anything. Also, the UV-light reduces in efficiency and should be replaced on a yearly basis, even if it's still radiating UV-rays. Ultraviolet light can harm the eyes, if the eyes are directly exposed to it. So, be very careful not to look at the UV-light directly. After every growth cycle, the pipes of the recirculation system should be cleaned inside and outside. Therefore, the pipes should have some fittings which can be easily disconnected. Some parts inside the tank shouldn't be glued at all. Pipe brushes with a long pole can be helpful.

Any time you empty the system, the cleaning should start immediately, because if the dirt dries up on the surfaces of the tank and other components, it becomes very hard to get rid of. The filter media should be cleaned as well. A high pressure cleaner works very well for this purpose. After cleaning, you can start another growth cycle by putting a new set of fish in the tank. If you want to do another spawning in the cleaned system, it's also advisable to disinfect with chlorine or hydrogen peroxide. After thoroughly rinsing, the system is ready for another spawning. Because the filter media will not have any bacteria yet, keep in mind that the recirculation system has to start up afresh.

The disinfection procedure is discussed in chapter 8 on reproduction, because in hatcheries disinfection is very important in order to get continuously good results.

Power supply for a recirculation system in a hatchery

The fish farm has to be run with constant light. Constant light means 24/7, without any interval of more than ten minutes.

Recirculation systems for hatchery operations do not really consume a lot of power when they are designed well. For example, a unit which can produce 120,000 fingerlings in a year or 40,000 juveniles in a year, can run on a power consumption of 100 watts per hour. The 100 watts are for the pump and the UV-light only. The consumption on a yearly basis is 900 kWh a year. The commercial rate for 1 kWh is approximately ₦ 25 and if the PHCN (Power Holding Company of Nigeria) provides power constantly, it will not be more than ₦ 22,500 per year to run one system.

With the current prices of fingerlings at ₦ 15/piece and ₦ 25/piece the output is respectively ₦ 1,800,000 and ₦ 1,000,000. In percentage, the electricity bill is not more than 2% of the sales price.

However, in Nigeria we cannot rely on a constant power supply. Therefore, fish farms need a back-up with alternative sources of power, which increase the price of running a system on electricity.

What are the alternative sources of power?

- Generator
- Battery pack with inverter
- Battery pack with inverter and solar panels

The cost of running a generator is almost three times higher than getting the electricity from the PHCN. It is not only the diesel or petrol consumption, but also the maintenance needed to keep the generator working all the time, which adds seriously to the cost of producing electricity.

I believe that in order to run a hatchery with recirculation systems, you need all the sources of electricity available:

- PHCN
- Generator (Figure 11)
- Battery pack with inverter and solar panels (Figure 12)

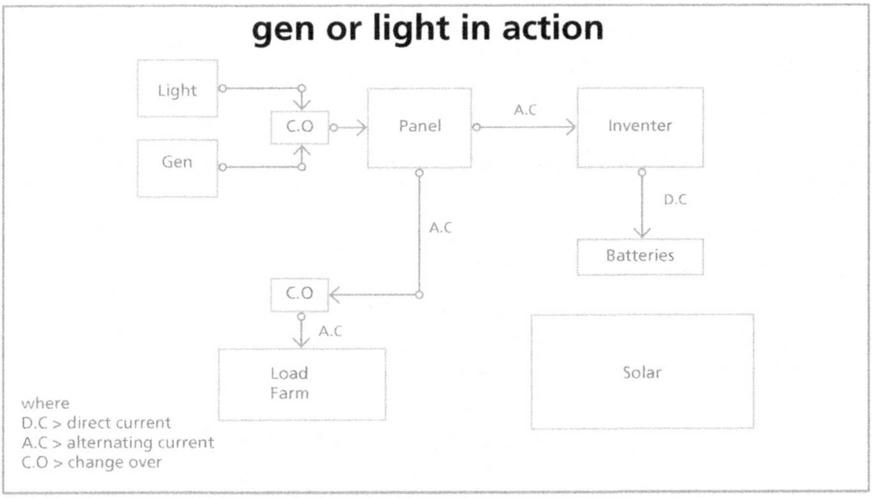

Figure 11. A generator or light (PHCN-power) in action.

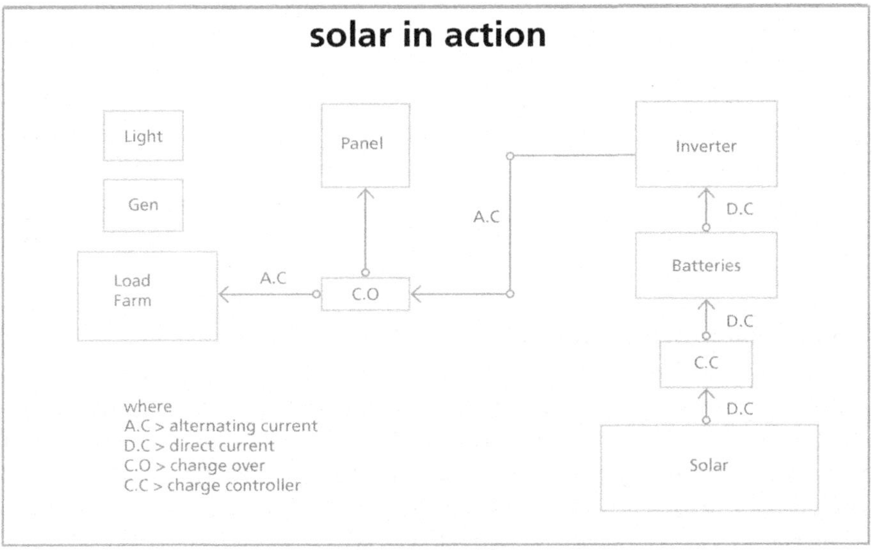

Figure 12. Solar power in action.

In a normal situation when PHCN-power is available for four to five hours a day and solar power from an average sunny day is coming in, the generator will still run for two to three hours, mainly during the night and early morning. Don't be deceived, when they offer you a standalone system for solar, and promise you will always have power from the solar system;

that is not going to happen. Sooner or later you will have no electricity at all and will be desperately looking for a generator:

- The weather situation can be bad for a couple of days, especially during the rainy season.
- The cables which connect the solar panels together, can become faulty by melting or partial contact.
- The cable connection between the batteries can become bad.
- Problems with the inverter occur.
- During the rainy season the PHCN-supply is less stable.
- One of the batteries gets bad and, if they have been placed in series, all the batteries of that battery bank cannot supply current.

The solar system should provide electricity for the recirculation systems and the borehole only. The borehole system should be switched on during the hours the sun is shining brightly and the batteries have already been charged well enough, when there is power from the PHCN or when the generator is working.

The borehole can have a solar driven pump installed. When the output of the pump is high enough to get enough water during the day time to also supply for the night, then it is a good option. The capacity of the reservoir has to be calculated and should be big enough to store water for 16 hours consumption. When there is not enough sunlight available, complications and shortage of water can occur because the pump will only work on DC and not on AC. So, a generator or electricity from the PHCN cannot be used to power it, because they provide AC. In this case, it's advisable to have two boreholes available, one on solar power and one on conventional power as a standby for the worst-case scenario.

About solar power in general

Solar systems that have been installed well (Photos 11 and 12), are relatively maintenance free and reliable. The shelf life of the various components varies from twenty years for the solar panel, ten years for the inverter to five years on average for the battery pack. The shelf life of the battery pack depends on how it is used and where it is placed; batteries should be installed in a ventilated cool, dry area. The inverter too should not be placed in the hatchery building, where it is hot and humid, but in a well-ventilated room. It is always advisable to place the inverter and the battery pack as close together as possible.
The solar panels should be put on the roof, with more than 5 cm space between the roof and the panels, in order not to accumulate heat. If the panels are too close to the roof, they get very hot. As a result, the cables can melt and the efficiency and life span of the solar panels decrease. In many cases the solar panels are first of all placed on scaffolding and not straight onto the roof.

Buy your solar system from a company with a proven track record. It is also advisable to buy it from a dealer who is close to your site, so that after sales service can be offered in time and when needed.
Always connect earth to the solar system and put an arrester on the roof, so that lightning will not affect the installation. The inverter is very vulnerable to power surges.
Buy maintenance-free deep cycle batteries. These batteries differ from normal car batteries in the property that they can be drained completely without being spoiled. The number of batteries should be enough to store the energy that is needed to overcome the period in the evening, night and early morning. The size of the battery is expressed in Ah. When the battery is 12 volts, the capacity of the battery expressed in Ah can be multiplied with the voltage (V) to yield the capacity in watts (W).

Example 5. How to calculate the required solar system capacity?
Suppose you have 16 batteries of 200 Ah and the voltage is 12 V. Formula: P = V x A. Total capacity in watts: 16 x 200 x 12 = 38,400 W or 38.4 kWh.
Assume the consumption of the farm is 1.5 kWh, you can run this farm on battery for circa 20 hours (30 kWh), because it is not advisable to run the battery totally flat. Normally the inverter will shut down when the battery drops below 20% capacity.
Assume you have a solar system and the farm takes 1.5 kWh 24/7.
The total consumption in a day is 1.5 kWh x 24 = 36 kWh. The solar system works approximately 8 hours on full capacity and should be able to produce 12 kWh in this time, in the meantime being able to store 24 kWh in the batteries. So, the capacity of the solar system should be approximately 5 kW per hour in order to take care of the farm 24/7.

Depending on the farm's power requirement, you need to have the appropriate size of inverters, the correct number of batteries and the correct number of solar panels. The golden rule is that the current through the wires should not exceed 100 A. If the current gets too high, it will burn the cable, which will be dangerous and expensive to replace. So, the cables would be thick. The only solution is to put the batteries in series (Figure 13). This will increase the voltage of the output, but not the amperage. That's why you will see the inverters using a higher voltage, when the capacity increases.

- Capacity of inventer: 1000 W = 83 A x 12 V
- Capacity of inventer: 2000 W = 83 A x 24 V
- Capacity of inventer: 4000 W = 83 A x 48 V

So, an inverter with a capacity of 4000 W will usually run on 48 V. 4 batteries of 12 V are put in series. More modules of 4 batteries are put parallel. One module of several batteries connected in series, is called a 'bank'.

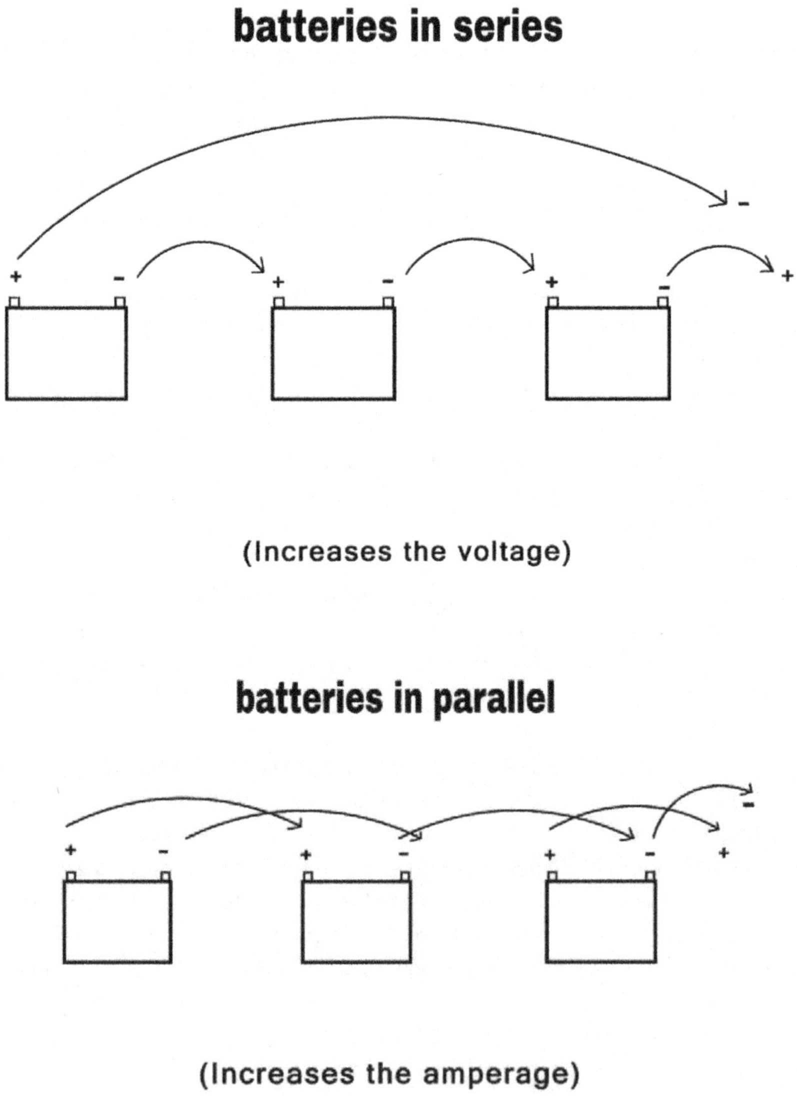

Figure 13. Batteries in series and parallel.

Photos 11 and 12. Solar panels, batteries and inverter.

4.3 CAGE SYSTEMS

Catfish can also be farmed in net cages floating in the water, or net cages attached to the bottom of the pond or lake. The difference is clear. When the water is very deep, the only way to put cages in the water is by allowing them to float at the surface with floating devices. The net will hang in the water with some weights on the edges, in order to create a cage where the fish can swim freely. If the level of the water is less than 1.2 m, you can also push sticks into the bottom of the pond or lake and attach the net along-side. These cages are also called pen cages. Traditionally they have no bottom net, only side nets. I consider this type of cage unsuitable for catfish, because the fish can dig in the mud and escape. In any situation the net should be enclosed on all sides except for the top. On Photo 13 you can see a module of 10 floating cages of 6 x 6 x 2 m (length x width x height).

Photo 13. A cage farm in a stone pitched reservoir.

Floating cages

How to avoid a floating cage drifting away?

There are two ways to get the cages fixed:

1. Anchors on the bottom of the pond, which are connected to the cages with rope.
2. The cages are connected to the shore with rope.

The first connection is the most difficult one, because the anchoring has to be done according to the size and number of the cages, the weather conditions and the current of the water in the dam or the lake. The best place for cages is where the optimal water exchange occurs and the water is very deep. In the case of a dam, the best place for putting cages is where the river bed was situated originally. This is the deepest place in the dam anyway. If the dam has fluctuating water levels, anchoring has to be adjustable. Anchors are more efficient at holding the cage structure in place if the bottom is soft and the anchor can dig itself into the soil. An alternative is to use concrete blocks. If the anchoring is too light, the cages will start drifting away from the initial place. In a worst-case scenario, you will end up losing the whole farm. Calculating the size or weight of the anchoring system can only be done by engineers with experience in the field of cage farming.

If the cages are situated not far from the shore, it is possible to hold the cages in position with ropes which lead to the shore and are hooked onto trees or concrete pillars. The cages can be accessed by a jetty. This construction makes it easy to harvest and feed the fish, because the cages are situated close by the shore.

Feeding floating feed in a cage?

Floating feed is advisable when cages are used. Sinking feed which is not directly eaten, will just sink through the bottom of the net. This is an immediate loss. Floating feed however, can drift away through the sides of the cages, if no proper care is taken. Therefore, each net requires a feed collar. This is a small mesh size net with a width of 40 cm which is attached to the complete periphery of the cage at the surface of the water. This way the feed cannot drift away to open water (Figure 14).

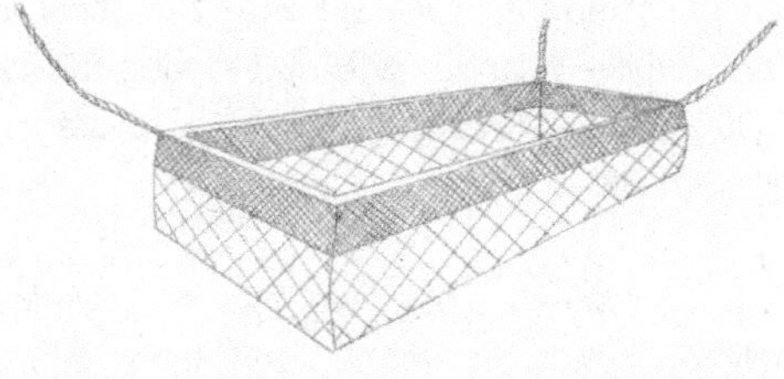

Figure 14. A feed collar to avoid floating feed to drift away.

Size of fish and stocking density in floating cage

Catfish should not be stocked in cages when they are very small. Their size should be at least 30 g. The juveniles first have to grow in ponds or concrete tanks to a weight of 30 g, then they are graded and transported to the cages. Small juveniles stocked in a cage, need a very small mesh size in order not to escape. But when they grow, the small mesh size will not allow a good water exchange through the cage. Dirt will also clog the small mesh size net and eventually the net gets very heavy and difficult to handle. The diameter of the net mesh should be 1.5 to 2 cm. Furthermore, juveniles tend to get affected by bacterial diseases in cages when they are too small. In addition, the net should not have a depth of more than 2 m, because if the catfish start swimming up and down through the water column, they will use a lot of energy, thus the feed conversion rate will not be optimal. Densities can be as high as 100 pieces per m^3.

Cage design and construction

If the nets or cages are placed in open water, for example a lake or dam, it should be clear that any damage to the nets or cages results in immediate loss of fish, without a chance of recovering them. Net quality is therefore of utmost importance; it should be thick enough, UV-resistant, well maintained and kept in a clean environment, because rats like to sleep and nest in it and make gaps. The specifications of nets are given in weight per m^2, thickness of the thread and mesh size. The nets should be knotless, so that they will not damage the bodies of the fish during the harvest. Any time a net is placed in the cage, it has to be thoroughly checked for gaps. Nets are made of several materials, either nylon or polyethylene is most commonly used.

Nets are vulnerable at their corners, especially when weights are put there which are not supported by a rope but instead are hanging on the net itself. In rough weather conditions nets can be shaken and tear apart after a period of time, resulting in large gaps.

If nets are placed in a big pond and the pond can be drained, escaped catfish are only a temporary loss, because they can always be caught again.

Cages can be built in any size. It depends on how many fish you want to sell in a year and your marketing strategy. A good size for example is a cage of 6 x 6 x 2 m, especially when galvanised pipes are used, because they have a length of 6 m. Stocking this cage with 100 pieces per m^3 results in about 7000 pieces. Growing them to 1.5 kg average, you can expect a harvest of 10,000 kg per growth cycle.

Cages in Nigeria are often built with galvanised pipes; two lines of galvanised pipes are welded together approximately 45 cm apart. In between these pipes, plastic drums are placed to create a floating platform. The nets can be hanged into the inner square of 6 x 6 m. The outer square is around 6.4 x 6.4 m. The plastic drums are usually 200 l. Check them thoroughly for leakages and seal the covers carefully, so that water cannot enter. For a cage of 6 x 6 m, use two drums on each side, making it eight drums altogether. The drums are lashed with rope between the galvanised pipes. Use good quality marine rope which is UV-resistant. The rope which you buy on the side of the road will not last longer than ten months.

Also, other options for making cages are available; there is one Nigerian company that produces floating elements of good quality plastic which can be mounted together. The nets can be placed inside them. It is more expensive, but more durable.

If you want to start a megaproject in cage farming, you can also contact overseas companies. They can give you turnkey solutions, including the anchoring system, predator nets, harvesting devices, ponton boats etcetera. Off course, buying from these companies comes at a price.

Cage farming of catfish is feasible, although currently you mainly see cage farms producing tilapia in Ogun and Oyo State at the dams.

Predators

Putting cages in a natural environment, attracts a lot of fish-eating birds and if not taken care of, this will certainly lead to severe losses. The most common birds are the kingfisher, the heron, the hammerhead bird and the cormorant. Within days of stocking, they will spot your site and immediately start to hunt; even when you have not seen any of those birds when building the project. The evidence, apart from seeing them on the cages, are the faeces, feathers and wounded fish. A bird net to protect your fish on top of the cage is obligatory.

Less obvious are the predators in the water, for instance monitor lizards, when cages are close to shore, crocodiles and tiger fish. If they enter the cages, they have already damaged the nets. The loss mainly consists of fish that have escaped through the gaps in the nets, because more fish escape than the predators actually eat. To overcome this problem, you can put an extra strong net around the net in which the fish are contained. This net is called a predator net, which is a very strong coarse 3 to 5 cm diameter netting. The net cannot be damaged by crocodiles or any other animal with sharp teeth. An extra advantage is that fish which escape from the first net, will still be trapped in the second net as long as they are big enough. Between the first net and the predator net a population of young wild fish can grow and eventually not escape the predator net anymore because of their bigger size by then. These fish are attracted and grow in this environment because of the faeces of the catfish, uneaten feed and the benthos living on the nets.

Applying for a license

When you start a cage farm in a dam, you have to apply for a license from the water authorities of the federal government who are in charge of the dam. After you have been handed over the necessary documents, you will usually get a license for a certain area of water defined by coordinates, where you can start your operations. The security of the operations is not the responsibility of the government, but has to be provided for by you. In order to get a good consensus with the people who live around the dam, you should visit them, explain your activities and tell them about the benefits of your presence. Employ people (ex-fishermen) from the communities around the lake. Mark the area, which is meant for the cage farm with floater balls, so that people know that this is a no-go area for them.

The security of the farm can be a big problem, especially if the operations of the farm are small and cannot pay off for the expenses made to secure the place. Most cage farms in Nigeria have a respectable size and, apart from having cages on the water, they also have land-based activities, such as hatcheries, nursery ponds and accommodations.

Harvesting catfish from floating cages

Harvesting the catfish from the cages is not so easy, especially if the construction aside from the cage is small and cheaply built with small planks. In principle harvesting consists of just lifting the net and crowding the fish together in one corner, scooping them up with a basket or catcher and carrying them to the shore or to the boat, depending on whether the cage farm is situated at the shore with a jetty or is a standalone cage in open water. If the nets have weights in the corners, they should be hung on a rope and taken out before the net is lifted. A long bamboo stick is helpful to lift the net to one side in order to reduce the volume of the net gradually, while shifting the bamboo stick from left to right. When the nets are crowded enough, you can start to catch the fish. When you want to take a break from catching the fish, you lower the net so that the fish are able to swim freely again and are not exposed to the sun.

Maintenance of floating cages

Maintenance of the nets is required to make sure they don't contain any gaps. After each growth cycle you can dry the net and clean it, if excessive algae or other debris are attached. The shelf life of a net depends on the quality, but after five to seven years the nets should be replaced for new ones. Losing a batch of fish because the net has gaps, costs more than replacing the net for a new one. Nets outside the water should be hung and dried in a dry area not exposed to sunlight and free from rodents.

Security of floating cages

As already mentioned before, when the area you are farming consists of a cage in open water, it is not very secure. Because there is no fence, it is accessible to anyone. In the day time security will not be so difficult, since you are operating the place and your presence is visible to everyone, and you see everyone too. The night however needs special attention. The place is dark and open. In order to secure the place, you can work with solar lights, sensitivity cameras and security guards on the water or on the jetty. Apart from theft of fish, the nets can be vandalised, which brings you even more losses when the fish escape. In my opinion, cage farming in open water is a business for well organised companies with good structures on the ground and on a scale, which allows good investments and spending money on security operations.

Water quality in dams and lakes

The water quality of dams fluctuates with the seasons. During the rainy season the levels of nutrients reduce but the suspended solid level increases, which makes the water more turbid. The water temperature also varies with the seasons. It is lowest during harmattan and highest during the dry season. The fluctuations are not that drastic for the fish and both in dry and rainy seasons there is usually no problem with the water quality. There is however one exception, which happens occasionally, once a year. The water from the bottom of the dam flows upwards, which results in mixing of water layers from the bottom and the surface of the dam. This normally doesn't happen. This process causes bad odours from hydrogen sulphide and methane gases and depletion of oxygen at the surface of the water for some days. Tilapia especially are sensitive to this situation and it can cause increased mortality. Usually after some days the mixing process stabilises; at that point even the natural fish in the water show signs of intoxication, although they have the opportunity to swim away.

4.4 EXTENSIVE POND CULTURES

In chapter 5 I mention the principle of pond constructing and the different materials which can be used to build one.
The difference between the flow-through system usually concrete tanks, and the extensive pond system, is the size at which they operate and their stocking density.

Ponds: 1000 - 5000 m^2, stocking density 5 - 10 pieces/m^2 preferably earth pond.
Tanks: 50 - 100 m^2, stocking density 50 - 100 pieces/m^2 preferably concrete tank.

Ponds, especially earthen ponds which consists of clay, tend to be large-sized with a surface area of 1000 m^2 or more. They are located in rural areas where land is abundant, and a sufficient water source is able to keep the ponds filled, even during the dry season.

Flushing is restricted to keep the ponds filled, because the release of nutrients by the fish is neutralized by the algae in the water and the binding activity of the soil. In other words, the ponds, if run extensively, can take care of the waste. Therefore, only at the end of the growth cycle, extra flushing is advisable.

Due to evaporation and seepage, the water consumption can be as high as 100 - 200 m^3 per hectare per day during the dry season. Catfish do well in this natural environment. The mortality rate is usually low, but the level of predation by birds, monitor lizards and other animals is higher than in intensive systems. Feed conversion rates tend to be slightly higher in earthen ponds too, because of the increased ability of the fish to swim around, which takes more energy. Hybrid catfish do very well in extensive earthen ponds and appear to adapt more to this natural environment than the pure Clarias gariepinus strain.

4.5 BACKYARD FARMING

Backyard farming is not really an alternative farming system, but it's worth emphasizing the principle and the advantages. Each of the farming systems mentioned before, can be used, but in almost all cases the backyard farm that works with these systems, uses the concrete pond for the flow-through system.

Backyard farms are small-scale farms situated in the garden of a house. The business is run by the family members, security, staff etcetera. The beauty of the backyard farm is the combination of interest; the water and electricity supplies support the house as well as the fish farm. The security and people are already there and they can easily take care of a small-scale fish farm. So, you don't have extra costs, apart from buying feed and fingerlings. The size of the farm is small and you wish to sell the fish to retail not all at once. This is very possible if you live in a residential area. You can even add value to the fish by smoking them. This way sales can be done any time you want, because the fish have been preserved. A backyard farm is also very good as a starting point for learning the ins and outs of fish farming. These provide for a basis for setting up a farm later on in a larger and more professional way in a designated area where there is enough water and land.

The waste water of a backyard farm can be a problem, if there isn't a good drainage system in the residential area. The water, in particular the water with solid waste, can be used in the garden, but the majority of the water will have to go out of the compound into the gutters. If the waste water remains around the compound, it will start to smell and attract mosquitoes and rodents, and your neighbours will start complaining!

Starting a small business is the safest way and it will give you time to learn the business without spending all your money. Mistakes can be corrected and are affordable. Learning fish farming can only be done by doing it. Your income should not depend on it when you are a newcomer in the business.

There is nothing wrong with thinking big, but start small. As a matter of fact, many entrepreneurs who start big, fail and turn away disappointed and broke from the business of fish farming.

5

CHOICE OF MATERIALS FOR FISH FARMS, SIZES, SHAPES AND NUMBERS

5.1 INTRODUCTION

A fish pond is defined as a water body which has an inlet and an outlet in order to manage the pond when it comes to draining, flushing and harvesting. If a pond does not have those features, it is not a fish pond in my opinion. In any case, a fish pond is not just a hole in the ground.

Building a pond and the choice of materials depend on many factors:

- Composition of the soil
- Size of the land
- Personal preference
- Capacity of the farm
- Financial capacity

5.2 EARTHEN POND

If the soil holds water, which occurs when the soil consists of a minimum of 30% of clay, you have the possibility of constructing an earthen pond (Photo 14). It is important that the dykes of the pond are well constructed and well compacted; they should not contain any stones or other materials, because they make the dyke permeable and weak. Usually the dykes of a pond have a slope of 30 - 45 degrees and a depth of 1.5 m. The water level should be 1.2 m which provides for a freeboard of 0.3 m. Small ponds of 100 – 300 m^2 can be excavated manually, but bigger ponds require machines for excavating and compacting. The top soil should not be mixed with the lower soil layers and should be placed back on the dykes and roads. Grass should be planted on top of the dykes in order to avoid erosion. Constructing an earthen pond should be done by experienced people. A surveyor has to be involved in order to get the levels correct. An earthen pond is, as I stated in the introduction of this chapter, definitely not just a hole in the ground. This book is not meant to be a manual for constructing an earthen pond, but it offers you an idea of the requirements for such a construction.

Many farmers use bags filled with soil to make dykes; I do not believe this is a durable solution. The bags certainly do not seal the dyke but they avoid erosion and reduce the seepage.

Earthen ponds are usually operated extensively. The stock of fish varies from 5 to 20 pieces per m^2. Earthen ponds need a lot of space and a lot of water due to evaporation and seepage. During the dry season evaporation can go up to 2 cm per day, which means 20 m^3 of water has to be added daily into a 1000 m^2 pond to keep the water level constant. Because in an earthen pond the soil and the water interact, the catfish really like this kind of pond. Their nick name is mud fish for a reason; they especially like the bottom of the pond. The only challenge of an earthen pond can be harvesting the fish, if the bottom of the pond is very muddy and the pond cannot be drained completely. Any fish left, will seriously interfere with the next stock of juveniles. The soil can be disinfected with lime after every drain or harvest and dried during the dry season. After some years, you will have to take out the excessive sludge. This sludge is perfect for improving the quality of any soil for farming crops, because it increases the organic matter in the soil.

Photo 14. An earthen pond.

5.3 PLASTIC LINED POND

If you are interested in building a natural pond, but the soil does not hold water, you can dig a pond in the same shape as an earthen pond and put a plastic lining in (Photo 15). For a plastic lined pond, it is important that the surface of the soil is smooth and does not contain any sharp objects which can puncture the lining. Also, the lining should be 1 mm thick and UV-resistant. UV-resistant means that sunlight doesn't spoil the lining and eventually cause it to crack after some time. The standard UV-resistant lining is not available in Nigeria, so it has to be imported. If the pond is big, the lining will be delivered in pieces and will have to be welded together. This is a job for a specialist with a special welding machine. If the welding is not carried out well, after some months or years the lining will just fall apart due to the detrimental effects of the wind and other weather conditions.

Photo 15. A plastic lined pond.

5.4 STONE PITCHED POND

If you want a natural looking pond, but the soil is not suitable and plastic lining is not available, you still have another alternative which is a stone pitched pond (Photo 16). In my opinion, stone pitching should only be carried out in such a way that the soil is well stabilized and erosion will not be a factor, otherwise it will undermine the stone pitching after some period of time. Stone pitched walls should be 12 cm thick and only be built by experienced masons in this field of expertise. Stone pitched ponds are not completely sealed, but that is not a problem because flushing and adding water is part of the water management anyway. The slope of the dykes is the same as for earthen and plastic liner ponds, which is 30 - 45 degrees.

A farm I visited, complained about unrecorded losses of fish from its stone pitched ponds. There were actually so many lost that the farm was not making any profit. Theft is always an excuse, but sometimes theft is not the reason and you have to look further into it. I found out that monitor lizards can be a serious threat if the environment is conducive for them. After one year, stone pitched ponds which are not built very strongly, will show openings between the stones or even worse, stones will fall out. This environment is ideal for monitor lizards to hide in. They dig a hole behind the stone pitching where they have shelter. In early morning and evening those lizards will be hunting your fish and you will actually have a lizard farm instead of a fish farm!

Photo 16. A stone pitched pond.

5.5 CONCRETE TANKS: FULL CONCRETE, BLOCKS OR COMBINED

In all situations it's possible to build tanks of concrete or blocks. This type of pond can be seen everywhere at small and large farms in remote or urban areas (Photo 17). If constructed well, they last forever and are easy to maintain. These ponds have vertical walls and harvesting the fish is slightly easier too.

The photo gallery in Appendix 4 shows you how to build a complete farm with reinforced blocks. In Appendix 5 a detailed bill of quantity for a full concrete tank versus a tank with reinforced blocks is offered.

The density of stocking depends on the flow rate of the water but can be as high as 100 pieces per m^2. Because of the costs of a concrete pond, the strategy is to farm as intensively as possible.

Photo 17. A concrete tank system.

5.6 PLASTIC TANK

In Nigeria, there are manufacturers that produce ready-to-use plastic tanks for fish culture (Photo 18). They are available in different sizes and usually have a circular shape. The volumes of the tanks vary from 500 to 5000 l. The tanks have a disadvantage though. Because the bottoms are not flat, harvesting fish from these tanks is difficult. To overcome this problem, you can put a layer of concrete on the bottom to cover and level it. This also makes the tanks very strong. The outlet should be positioned right at the bottom so that the tank can be drained till the last drop.

Plastic tanks are very good for the pre on-growing stage. At these stages the fish weigh between 1 and 50 g, which is between the hatchery and the grow-out stage (see Appendix 1 for the growth stages of catfish). For small scale farmers these tanks can also be used for grow-out.

Plastic tanks are circular because this shape makes them strong and most cost-effective, compared to square or rectangular tanks.

Photo 18. Plastic tanks for juvenile production.

5.7 FIBREGLASS OR POLYESTER TANK

Fibreglass or polyester tanks are not very common in Nigeria. But there is an industry that produces boats using the same material. On request a company is able to produce fish tanks too. As long as the tanks are made correctly, they are very durable and suitable for fish farming. The sizes are in the same range as for the plastic tanks between 500 to 5000 l, and can be rectangular or circular (Photo 19).

Photo 19. Fibreglass or polyester tanks.

5.8 SHAPE OF THE POND

You want to build a pond which is efficient and as cheap as possible. This can be expressed as costs of material per m^3 of the tank. More interesting to know is that with tanks of the same volume a rectangular shape is more expensive than the square one and a square tank is more expensive than a circular one. In addition, bigger sized tanks will be cheaper than smaller sized tank per unit of volume.

Compare the following tank sizes with the same volume:

20 x 5 x 1 m rectangular tank: 100 m^3 (length x width x height)
10 x 10 x 1 m square tank: 100 m^3 (length x width x height)
Diameter 11.2 circular tank: 100 m^3 (πr^2 x height)

Comparing the costs of building a tank is done by calculating the surface area of each tank, because the more material you use, the more expensive the tank will be, whether it is built in plastic, concrete, blocks or stone pitching.

20 x 5 x 1 m rectangular tank: 150 m^2
10 x 10 x 1 m square tank: 140 m^2
11.2 diameter circular tank: 135 m^2

The rectangular tank has 15 m^2 more material and is therefore 11% more expensive than a circular tank.

The advantages and disadvantages of circular, rectangular and square tanks

Circular tanks

Advantages

- Relatively cheap in material costs.
- Strong.
- Self-cleaning and no corners where waste or dead fish can accumulate.

Disadvantages

- More difficult to harvest the fish.
- Setting up a farm consumes a lot of space because the tanks have no mutual walls; all tanks stand on their own.

Rectangular and square tanks

Advantages

- Efficient in setting up the farm with little spacing or no spacing at all between the tanks.
- Easy to harvest compared to circular tanks because the fish can be concentrated into the corners.

Disadvantages

- More expensive in material costs.
- Accumulation of dead fish and waste in the corners if the flow rate is insufficient.

I estimate that in Nigeria 50% of fish production is done in earthen ponds, 40% in concrete ponds, 5% in plastic lined ponds and the remaining 5% in the other alternatives such as plastic tanks, fibreglass tanks etcetera.

Most farmers prefer rectangular tanks. This is a good choice, but for small tanks you often see the circular plastic tanks.

In Appendix 5 you can find a bill of quantity (BOQ) for a 10 x 5 x 1.3 m pond made out of reinforced blocks or full concrete with early 2019 prices.

5.9 POND SIZE AND NUMBER

When you are ready to start building ponds, you probably have already chosen the shape of the pond and the material you want to use, but you have not yet decided how big the ponds should be. How big should a pond be? For the on-growing stage of the growth cycle, the pond should be of a size which enables the stocked fish to be sold within a week when harvest is due. You also have to determine the number of sales days you want to have in a year. The more times you want to sell, the more ponds you need to build. It is advisable to have regular sales, for example on a weekly basis. This allows you to build a customer base with regular customers.

Example 6. How to calculate the number of ponds you need?
Suppose your production is 100 tons a year and you want to sell on a weekly basis. This means 52 sales moments or days in a year. Let's assume the growth cycle of the fish is 175 days; in this case the fish remain in the pond from stocking till harvest during 175 days. This means you can do 2 cycles in a year. Each pond has 2 harvests in a year. So, for sales on a weekly basis, you need 52/2 = 26 ponds! This means you stock on a weekly basis and you harvest on a weekly basis. If you want to sell on a monthly basis, you need 12/2 = 6 ponds.

Many new farmers are tempted to stock all the ponds at once. This will not work at all. You will get a serious problem when the fish are ready for harvest. There are four reasons:

- Your working capital to grow all the fish at once, is very high.
- You have to sell all the fish in a short period of time.
- Customers cannot buy regularly but only a couple of times a year.
- Your need for water is very high just before the sales whereas it drops to almost zero after the sales.

When the ponds are too big and the sales from one pond are spread over a longer period, let's say three weeks, the result will be bad. The fish from the first sales will be bigger than the fish sold after three weeks. Because during these three weeks feed will be given periodically, the fish will be severely stressed so that they eat without growing.

Appendixes 2 and 3 show how the maximum working capital increases with only two sales in a year compared to twelve sales in a year.

Another problem I notice is that many fish farms operate with different sizes of ponds. To manage this is very difficult, because all the ponds will have a different number of fish and the feeding regime has to be adapted for each single tank. Also, the sales quantities are not regular, they depend on the size of the pond that is harvested. If you want a smooth and easy operation, you need to build all the ponds the same size.

6

HOW TO SET UP A FISH FARM

Setting up a fish farm is not the goal, but just a tool in order to start farming fish. I have seen investors who love to build a farm, but when it comes to farming, they are nowhere to be seen. The joy is over!

When you have found the best location and have acquired the land, it's time to start. The first step to take, is making a survey of the land including all the levels of the land, where there is water and the distances involved.

With this survey you can start making the design and put all the facilities into place (Figure 15). The design can also be used to make an investment plan, a production plan and a profitability and cash flow projection. You will find more about the profitability and cash flow projections in chapter 12 on the economics of fish farming.

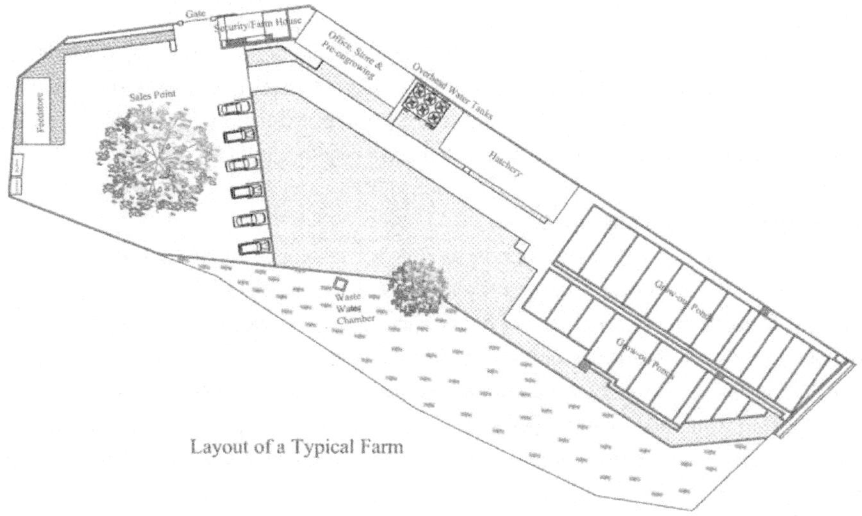

Figure 15. Layout of a fish farm.

A fish farm needs the following facilities:

Fence or wall

Fish farming is a capital-intensive business. When the ponds are not protected, you will certainly get theft problems. It is not only birds with feathers which can take your fish, also birds without feathers can do that! A fence or a wall can be constructed with blocks or it can be a wire fence. Both will serve the purpose. When the area is very swampy, it's advisable to make a fence out of wire. It is wise to build the enclosure before starting any other job.

Fish ponds

The fish ponds are usually located between the lowest and highest level of the land. The storage reservoir is normally situated at the highest level and the drainage at the lowest level. If the land does not have many differences of level, you must create levels and slopes, in order to have the correct flow of water. The ponds should be easily accessible by car, tractor or at least a wheelbarrow for a small farm in order to provide the feed and transport the fish when they are due to harvest. Most investors start by building ponds but instead they should first start with the roads and water infrastructure. Fish don't grow in a concrete pond; they grow in water. Without water the ponds are useless!

Power house

If PHCN-electricity is available, the generator house should be located close to where the PHCN-power comes into the farm. Since it is important that this spot is supervised, it should not be situated far away from the security house. Normally you will see the power house, security house and gate house located close together or close to the gate.

Gate house or security house

The gate house is the place where goods, services and people coming in or coming out, are recorded. It is very important for security to check the inflow and outflow, and be aware of what is going on. Another aspect of the gate house involves biosecurity. Don't allow people to enter the farm with contaminated materials. People should be directed to the person they want to see. They should not start walking around the farm unattended.

In bigger farms, it's advisable to give visitors a pass which clearly shows they are a visitor. Thus, the gate house plays an important part in keeping the environment safe and theft-proof!

Accommodation for the staff

Most investors do not give too much attention to accommodations for the staff. They come with costs and do not benefit the production. However, the farm will always have some key staff walking around during the night. If they are not comfortable, this will certainly affect their performance. So, if your ambition is to develop a big farm, you should also take in consideration that staff accommodation is a serious matter. Water, light, security, a bed, a kitchen, a shower and a toilet are the basics for workers staying 24/7 at the farm. Many investors have in mind to provide all those things when they start making money; but the sad story is, this way the money will never come!

Hatchery-, pre- and on-growing ponds

Depending on what you want to do and the size of the farm you wish to have, a hatchery and an outdoor facility to grow fingerlings and juveniles (Photo 24) to the stage of post-juveniles prior to stocking the outgrow facility, are important. See Appendix 1 for the growth stages of the fish.

Farms with the ambition to produce over 50 tons of fish each year, should consider having their own hatchery, but they will certainly need a facility to grow juveniles and fingerlings to post-juveniles. Farms that rely on supplies from other hatcheries for stocking fish, can become very disappointed in terms of quality and regular supply of the fish. This will result in periods without stocking and inevitably no supply after five to six months. A farm depending on an erratic supply of fingerlings cannot survive!

Water storage facility and boreholes

Water is the lifeline of the farm. After building a fence or wall, this should actually be the second action to be taken. When you have drilled a borehole and made the water storage facility and installed the supply line, this will also supply your needs when you build the remaining structures, because for building you need water! A survey of the land in order to get a good spot for drilling a borehole does not guarantee a steady supply for water and a good yield. The proof of the pudding is in the eating! After drilling and putting the borehole pump into place, you will find out the real capacity of it. This will help you to decide how many tanks you can build.

Intake from a river or dam

The intake is actually the heart of your farm; it is the supply line of water. The dimensions of the intake should be determined by the water needs of the farm. A river or dam contains normally more than enough water to supply the farm. Officially when you start a farm and you take more than 250 m^3 of water in a day, you need a license from the river basins development authorities, which operate at state level. A license can only be issued if you make an environmental assessment report. When you start small, you will get away with pumping water from a river or dam, but if you want to start a big farm, sooner or later this organisation will catch up with you. It is better to face the music before starting to dance!

Materials and feed storage

Storage facilities for materials and feed can be built with concrete blocks, but I prefer 20- or 40-foot containers (Photo 20). These containers are well protected against rodents and can be closed and easily protected from theft. You need to roof the container in order to keep the heat out. This is important when stocking feed. Usually you make a base to put the container on, so that flooding or rainwater will not disturb it. Storage of feed should be done in a strategic location for easy access.

Materials storage concerns materials that are used at the farm, such as graders, catchers, baskets, scales, drag nets etcetera. All these materials should be kept in a dry place. As far as the nets are concerned, it is important to store them in a rodent free area. Rats and mice especially like to make nests in nets, which leads to large holes if they are not taken good care of. When you leave your harvesting equipment close to the ponds, you make a tempting environment for theft. After use, harvesting equipment should be locked up at all times.

Photo 20. Feed or materials storage, made from 40-foot containers.

Administration block

When the farm is relatively small, all those facilities mentioned above, can be combined in a multifunctional building. When the farm is big, it's advisable to have an administration block. From this building purchases and sales are directed, cash payments are collected and petty cash can be received. In the administration block you can find a cashier, an accountant, a marketing or sales manager, the staff manager and the store keeper. But if the farm is small it will be just one person doing everything. So, a small farm needs a general manager who understands all the disciplines and is able to handle them.

Point of sale

When the farm is small, you will most likely want to sell your fish close to the ponds, because that is where the fish are harvested. But this is not an ideal situation, because a point of sale near the ponds will disturb the fish in the other ponds. Also, for biosecurity reasons it is not advisable to allow buyers to come near to the ponds with their dirty baskets.

Selling fish in all situations, should be carried out under shade conditions with no exposure to sun, with access to clean cold water supplied by a hose from a tanker or storage tank, and a smooth floor that will not damage the skin of the fish when they accidentally fall out of the basket (Photo 21). The simplest way is to put a plastic lining on the floor where you weigh the fish and sell them. Ideally, there should be drainage to allow excess water to flow out.
If you want to make a point of sale at your farm, it should be close to the gate and the administration block with easy access for customers. Consequently, the fish harvested from the ponds, should be transported to the point of sale.

Photo 21. A point of sale at a fish farm.

Internal roads and drainage system

To fast track operations, you will need internal roads and drainage (Photos 22 and 23). The easiest way to make a road is with gravel and capstone. On the blueprint of the farm it should be clear where roads are needed. The levels on the land will show you where the gutters and drainage systems are needed in order to release excess water either from the farm or from the rain.

Neglecting roads and drainage will result in a muddy, dirty place, which makes operating smoothly difficult.

Photos 22 and 23. Internal roads and a drainage system covered with planks and a security post at the back.

Photo 24. A pre- and on-growing facility with two different supply lines for water: one from a dam and the other from a borehole.

7

FEED AND FEEDING

1. THE GROWTH CURVE OF CATFISH
2. THE GROWTH RATE OF CATFISH
3. FEED QUALITY
4. IS IT ADVISABLE TO MAKE YOUR OWN FEED?
5. FEEDING CATFISH
6. FEED CONVERSION RATE AND THE ECONOMICS OF FEED
7. FEED STORAGE

Some years ago, I went to a big farm. I was asked to analyse the farm, since the profit they were making was not encouraging at all. They wanted to know what actually was going on. So, I investigated the farm by moving around with all the staff for a couple of days without interfering in their behaviour and handling. Feed is a crucial aspect in the economics of fish farming, since it is by far the highest cost factor. I noticed that the people dealing with feed, were not experienced. Actually, neither the farm's supervisor nor any of the senior staff were involved in the feeding. They left the feeding to IT-students, day laborers and low-rated workers without supervision. So, the feeding was carried out badly without any vision and any idea of how to grow fish. The senior workers remained in the office, equipped with air conditioning and gossiping the entire day. Feeding fish is the main job of the day and should always be supervised by the most experienced staff. At feeding time, you interact with your fish. This provides information on how the fish are doing in relation to their health and growth. Never allow feeding to be carried out by inexperienced and unmotivated workers!

7.1 THE GROWTH CURVE OF CATFISH

The growth curve of most animals has an S-shape; catfish are no exception. Figure 16 shows a slow growth in the first two months: from 0.002 g as a fry to 5 g as a juvenile (see Appendix 1 for catfish growth stages). From this stage the growth accelerates till they reach 1 kg in a period of four months. From 1 kg onwards, the growth rate declines, reaching to almost zero when the maximum end weight of 3 to 4 kg is reached.

Catfish actually never stop growing. Depending on environmental factors and feed, they keep on growing, but only very slowly after a certain point. Brood stock which is well fed and taken care of, can reach up to 8 kg in two years!
The growth curve of catfish depends on the feed quality, the quality of the environment, the genetics of the catfish strain and the water temperature. The growth curve shown in Figure 16 is based on an ambient temperature of 25 - 27 °C and a high-quality extruded feed with a protein content of at least 45% at the beginning of the growth curve and 40% once they reach the adult size.
The growth curve shows an average of the total population of the fish. If the individual growth of the catfish is used, the growth curve will show an enormous variation.

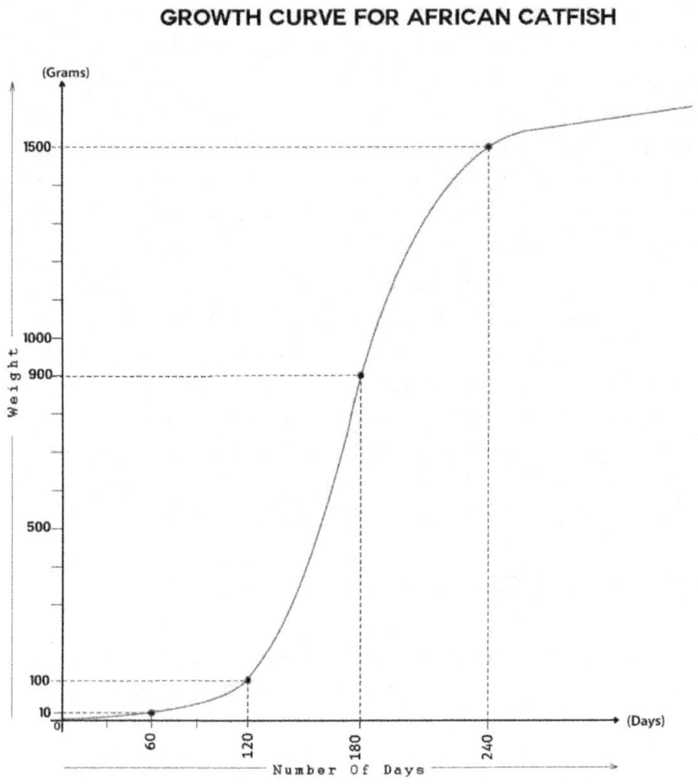

Figure 16. The growth curve for the African catfish.

As fish farmers, we are not necessarily interested in individual growth rate figures, we want to know the average growth rate. When a farmer tells you that fish can reach 1 kg in three months, he is most likely talking about a shooter that has reached this weight at the expense of other fish. When it comes to selecting brood stock, it can be useful to select the fast-growing fish, but not the shooters. They are cannibals.

The range of growth is shown in Figures 17 and 18 and is based on the Gaussian distribution. The chart in Figure 17 shows an ideal range of growth. The chart in Figure 18 shows the range of growth in a batch of fish with a lot of growth variation. This implies that the farmer is less skilful in reaching his target of 1.5 kg.

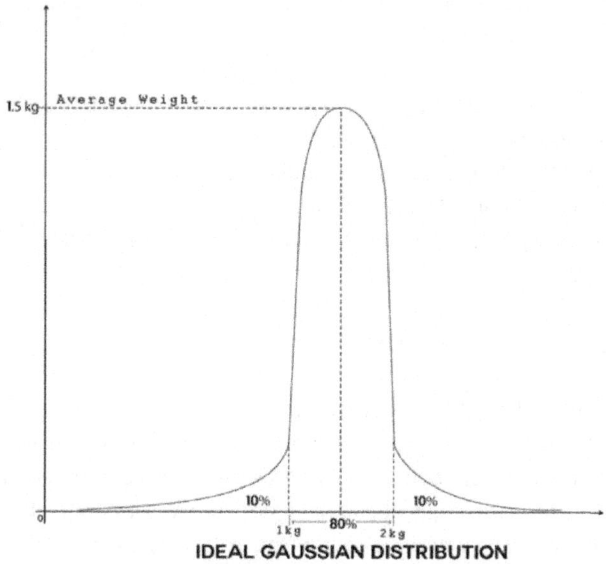

Figure 17. The Ideal Gaussian distribution in a batch of catfish. The horizontal axis represents the weight of the fish, the vertical axis represents the number of fish in the population with a certain weight.

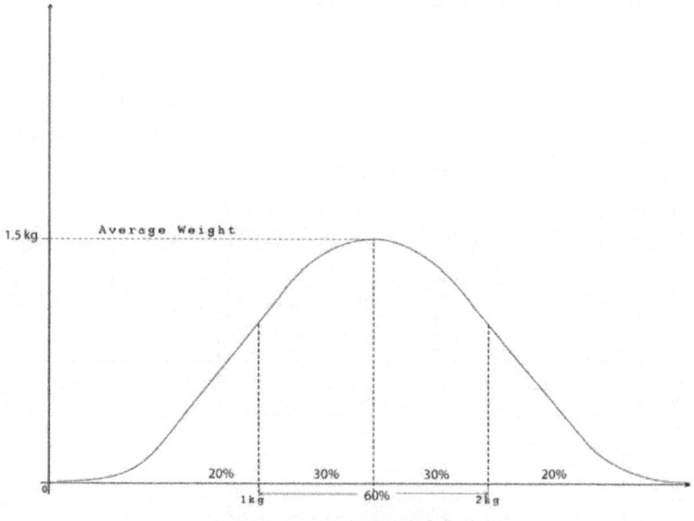

Figure 18. A less optimal Gaussian distribution in a batch of catfish. The horizontal axis represents the weight of the fish, the vertical axis represents the number of fish in the population with a certain weight.

What you are interested in is a batch of fish with little variation in sizes. When the sizes vary a lot, it indicates the following:

- Grading fish before stocking them has been done badly: the runts and shooters were not taken out.
- Disease problems have occurred during the growth cycle.
- Feeding has been carried out erratically.
- The genetics of the stocked fish originated from different batches and were of poor quality.

This results in a poor survival rate and reduced profit.

How to minimize the growth range?

You can minimize the growth range of the fish by grading them very uniformly, before final stocking into the grow-out pond. The juveniles should all be from the same mother and father (full siblings!). The moment you have stocked juveniles from different genetic backgrounds, there will be growth variation resulting in cannibalism, feed competition and higher feed conversion rates.

7.2 THE GROWTH RATE OF CATFISH

The growth rate of catfish depends on their size; the smaller they are, the higher the growth rate. Look at the curves in Figures 17 and 18 and the table in Figure 19 with the growth rates from 0.5 g to 1.3 kg on good quality feed and optimum water temperature. It takes six months to grow a fish from 0.5 g to 1.3 kg.

Growth (g)	Days	Growth rate (%)	Feeding rate (%)
0.5 - 1.0	6	12	6
1.0 - 2.0	6	11	6
2.0 - 5.0	9	10	6
5.0 - 10	9	8	5
10 - 20	9	8	5
20 - 40	10	7	5
40 - 80	12	6	4.5
80 - 160	14	5	4
160 - 320	20	3.5	3
320 - 640	30	2.5	2.5
640 - 1280	60	1.2	2-1.2
	6 months period		

Figure 19. Growth rate in relation to the weight of the catfish.

7.3 FEED QUALITY

A definition of a good quality fish feed is a feed that meets all the nutritional requirements of the fish. So, the first thing you have to do is find out what they are. Luckily this has been done in many studies and experiments at scientific and practical level. It is important to know that a catfish is basically carnivorous, which makes the requirements for protein, in particular animal protein, relatively high. The digestive system is short and any ingredient which is not easily digestible, will be less beneficial for their growth.

So, making a good quality catfish feed, is easy in principle. The challenge however, is to make a good quality catfish feed at a competitive price! Prices of fish feed ingredients change almost on a daily basis; so, the formula for the feed is not static but a continuous process of adjusting it to the low-cost formulation without reducing the quality of the feed. Prices of ingredients are changing constantly due to demand and availability. The availability changes when the harvest of important ingredients, such as soya, goes up or down. The availability of fish meal depends on climate factors; it is a product which is harvested from natural resources and limited available. Therefore, it is subject to large price fluctuations.

Some ingredients contain essential nutritional factors. Without these ingredients the fish feed will not be complete. This will result in fish developing problems like ceasing growth and/or getting sick. For example, the essential amino acids in proteins cannot be replaced by any other amino acid. The same is true for vitamins and minerals. If the feed doesn't provide the basic requirements for these nutrients, the fish will certainly become sick and die.

Fish feed can be seen as a chain of ingredients which meet the requirements of the fish. If one link in the chain is not sufficient, the chain will be as weak as the weakest link.

Making your own feed is not something you can do overnight!

Besides the feed formula, the appearance of the feed influences the quality and performance of the feed too.

Uniformity

Feed is produced in different sizes. Depending on the size of the fish, the feed size will be bigger or smaller. The range of feed is shown in Appendix 6. High quality feed has uniform particle sizes which all have the same appearance. Also, you cannot distinguish the ingredients inside the particles, such as corn, because they have been ground finely. Well ground feed is more stable in water and is easier to digest.

Water stability

The feed is given to the fish in the water and, if it is not eaten immediately, it can disintegrate and is wasted. Starter feeds, especially those which do not float, should be water stable for some time. The choice of ingredients and the machines used to produce the pellets, make feed water stable.

Palatability

The feed should be tasty for the fish. This can be achieved with ingredients such as blood meal and fish oil. Special products can be added to the formula in order to improve the taste even more. As a result, the fish will eat more feed more quickly and the growth rate will improve.

Dust free

The feed should be dust free. Feed factories have a special machine to separate the dust from the feed particles. The dust is reclaimed in the process, so it's not wasted. However, if the feed does contain dust, it only pollutes the water and it doesn't benefit the fish at all. If the feed is not up to thorough handling in bags, for example carrying and piling up, it can cause a lot of dust.

Floating

If the feed can float, it will allow you to have a good view of the feeding behaviour of the catfish. You will immediately notice when the fish have had enough. This way overfeeding can be easily avoided. Floating feed usually is produced with extruder pelletizing machines. Another advantage of feed from this kind of machine, is that it is more digestible because it has been processed under high temperatures.

Expiry date

The expiry date of purchased feed should not be within three months. Feed millers can achieve this by drying the feed to at least 90% dry weight and adding antifungal additives.

7.4 IS IT ADVISABLE TO MAKE YOUR OWN FEED?

Many fish farmers think about making their own feed. They believe it is something they can do easily alongside farming the fish. In their opinion it is also very profitable because it reduces the production costs of the fish. All this is based on assumptions and gossip from other farmers. It's an idea born out of ignorance, because if you know all the ins and outs of feed milling, you will certainly think twice before starting a feed mill yourself.
Starting your own feed mill should only be considered if there is no alternative. Don't be a jack of all trades and master of none!

Nowadays Nigeria has at least four professional feed millers with the professional equipment, purchased from overseas and owned by multinationals based in Europe and the USA. Their scale of production is massive (2000 - 4000 tons per month) and efficient. They have good connections for purchasing raw materials at competitive prices and laboratory facilities to check the quality of the raw materials. They use software to change the feed formula based on the low-cost formulation without minimizing quality. Because quality control is carried out during the entire production process, the feed always has a consistent quality. The prices for feed are very competitive, because feed millers need this scale of production in order to survive. Net profit margins are less than 10%. They employ a lot of marketers and have a wide distribution network. Before starting your own feed mill at farm level, you have to ask yourself: "Can I compete with companies like this? And, is it not better to have a good agreement with a feed mill to purchase from them instead of producing it myself?"

Let's have a look at a professional feed mill and list all the stages which feed ingredients go through, before they become fish feed (Figure 20).

- Warehouse for stocking ingredients such as soya, corn, fish meal, concentrate, pre-mixes etcetera.
- Intake of feed ingredients; from a bunker to the hammer mill.
- From the hammer mill to the mixer.
- From the mixer to the extruder or pelletizing machine, where the ingredients are transformed into pre-feed.
- From the extruder to a dryer, where the pre-feed is dried.
- From the dryer to a screener to get rid of dust.
- From the screener to a coating machine where the feed possibly is coated with fish oil or soya oil.
- From the coating machine to the weighing and packaging machine.
- From the packaging machine to a warehouse for fish feed where it is ready to be sold.

Additional equipment and facilities

- Boiler to produce steam for extruding the feed or pelletizing the feed
- Laboratory for quality control
- Generator house and transformer for PHCN-supply (generator up to 500 kVA)
- Diesel tank
- Control panels
- Workshop
- Store for spare parts
- Warehouse for empty feed bags

- Cold store for sensitive additives such as vitamins and minerals
- Administration block
- Canteen
- Borehole(s), water storage tank and water conditioner

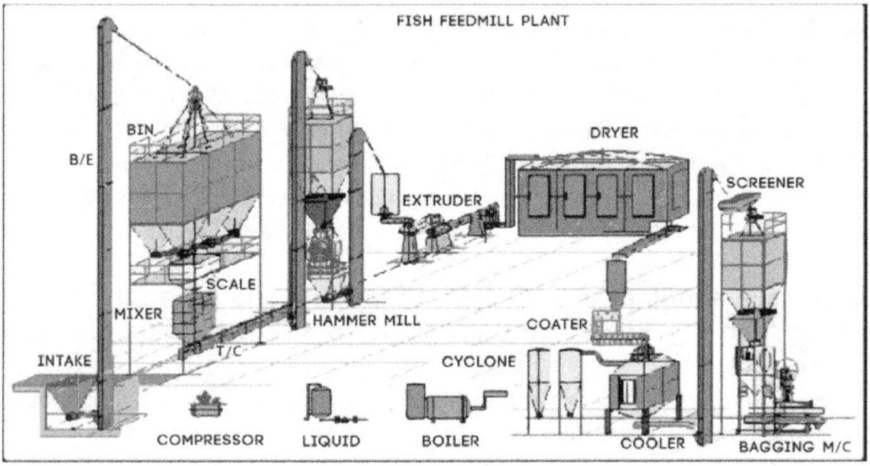

Figure 20. Schematic impression of a feed mill set-up (source: Muhammed Umar Aziz).

Apart from mega fish farms producing thousands of tons each year, it should be clear that this type of investment cannot be done at farm level.

7.5 FEEDING CATFISH

Feeding catfish is the most important part of the job. The feed cost can be as much as 75% of the cost price of the fish. So, feeding should be carried out by the most experienced staff who have a good eye for the fish and their behaviour. Only then can the amount of feed be adjusted to the appetite of the fish.

The following feeding levels can be distinguished:

No feeding

It's obvious: no feeding results in no growth, but that's not all. The fish will also decrease in weight with at least 1% every day. Insufficient feeding happens a lot among Nigerian fish farmers due to bad financial planning, not purchasing feed in time or lack of water.

One day without feed, results in two days without growth. It is clear that the economic results will be bad at the end of the cycle when the fish are harvested.

Maintenance feeding level

Just enough feed given to the fish to keep them at the same weight, is the maintenance feeding level. At this level there is no weight gain and no weight loss. This is a loss too, because the feed is wasted.

Underfeeding

Underfeeding happens a lot at Nigerian fish farms and is partly due to the difficulty in finding the optimum level for feeding. Some farmers are just afraid to overfeed the fish and as a result are underfeeding. Also, when the water supply is limited, farmers tend to underfeed to avoid pollution. Low availability of feed and its financial consequences can cause the farmer to start rationing the amount of feed. This is easily preventable though. Make a financial plan for the feed you need before you stock your fingerlings or juveniles (see Appendix 1 for the growth stages of catfish). If you don't have the financial means for the optimum feeding level, cut down the number of fish you want to grow. If the optimum feeding level is practised, this strategy provides for better economic results, despite the lower number of fish.

Optimum or maximum feeding level

The optimum or maximum feeding level is the level you have to obtain to maximise your profit. It is the quickest growth you can achieve with the fish you have stocked under your farming conditions. Only experienced fish farmers are able to achieve this level during the complete growth cycle.

Overfeeding

When you overfeed, you feed more than the fish can eat. If you use floating feed, this will be easy to see; the feed keeps floating at the surface of the water without being eaten. It will pollute the water and it is really a waste of money. Some farmers believe that, if the fish get hungry, they will eat it later. The truth is, catfish don't like to eat old feed floating at the surface. The taste has completely gone.

The different types of feeding levels are given in the diagram below (Figure 21). It also shows another parameter which is called the feed conversion rate or FCR. The meaning of the FCR is explained in paragraph 7.6.

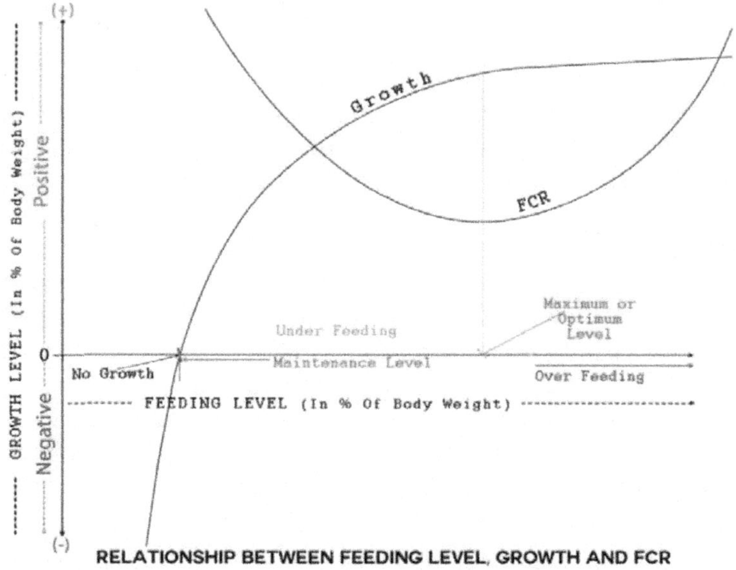

Figure 21. The relation between feeding level, growth and feed conversion rate (FCR).

How to feed?

Feeding the catfish should be done by experienced staff, because at feeding time you have contact and interaction with your fish. The way they eat, gives you information about their health, whether to increase or decrease the feed given the following day, about the uniformity of the fish and whether the fish are growing well. Normally, the feed given has to increase on a daily basis. Many farmers are reluctant to do this. Instead they keep the amount of feed constant for one week and increase it drastically the next week. If you study the feeding pattern of many farms, you will notice that they like to feed in numbers of bags. 1 bag is 15 kg, so it will be 15, 30, 45 kg a day, which means that the increase of feed is one full bag. So, the pattern will be like this: 15 kg for 1 week, an increase to 30 kg the next week for 1 day only, and the second day of next week back to 15 kg again or even less. It doesn't work this way. You should never increase with a full bag at once and if you do so, it will result in poor feed intake the next day. There is not a steady increase in the amount of feed given. This will certainly affect the growth efficiency and the conversion efficiency of feed into fish. Why farmers are doing it like this, is obvious: to feed full bags of feed, doesn't require weighing out the feed, so feeding can be carried out very quickly. It's laziness which makes them to do it like this.

If you want to grow catfish efficiently, you need to increase the amount of feed on a daily basis. The daily increase should be based on the feeding schedules provided by the feed companies. Appendix 7 shows you an example of a feeding schedule; it is a simplified version but it serves the purpose.
A feeding schedule however, is not more than advice; differences occur for different reasons, such as lower water temperatures during harmattan, disturbance of the fish, bad water quality or the fish are unwell. If you cannot finish the feed scheduled for a pond during the day, it has to be found out why this is happening. This calls for action. Reduce the level of feed to the amount they can eate and increase it from this point onwards. Never create big differences in the amount of feed; it should always be gradual.

The amount of feed in a feeding schedule is always provided as a percentage of the body weight. The percentage given to the fish varies and decreases from 6% for juveniles to a little above 1% for table-sized fish.

Example 7. How to weigh out the daily feeding for juveniles?
You have bought 2000 juveniles of 6 g average weight and you want to weigh out the daily feeding.

Check the feeding schedule in Figure 19 to see what the juveniles should get as a percentage of their bodyweight: the feeding advice for 6-g fish is 5% of their bodyweight.

Total biomass: 2000 x 0.006 kg = 12 kg

Feed given: 12 x 5/100 = 0.6 kg

The first day when you have stocked the fish in the pond, you will not give the full quantity of feed but only 50% of it. In 2 days, you should reach the level of 0.6 kg and increase from that point onwards.

Overfeeding of just stocked fish will increase the risk of outbreaks of disease. Keep them a bit hungry till they are used to the new environment.
When you start feeding a pond, start with a little bit to see the initial response of the fish, before feeding the remainder. Feed everywhere in the pond, so that all the fish have the opportunity to eat; especially when you add floating feed. Spot feeding is recommended when you feed the fish pellets which sink.

Frequency of feeding

The number of times a day you feed depends on the size of the fish. Smaller fish, like the hatchery stage, need feeding more frequently during the day than bigger fish, like the grow-out stage. Catfish under 20 g are fed between five to seven times a day. Fish above 20 g are fed three to five times a day.

The feed should be weighed out for the full day and put into a bucket which is placed near the pond or tank side. The bucket should have a cover so that rain cannot spoil the feed. For hygiene the feed is never given by hand, but with a spoon or a small bucket. At each feeding you keep in mind how many there are to go and distribute the feed equally over the day. If possible, feed in the evening too and also during the night. In some cases, lack of manpower will not allow evening or night feeding although it is advisable for fish in hatcheries to feed frequently.
If there is a shortage of manpower at the farm, you can decide to use feeding machines. Some examples of these machines are given in photos 25 and 26.

Disadvantages of feeding machines

It is assumed that the feed put into the feeding machine will be consumed; the machine will therefore not stop feeding, when the response of the fish is low. So, you still have to check the feeding behaviour of the fish and adjust the machine to it. An exception is the self-feeder. The self-feeder is very popular at catfish farms in Europe. These machines need maintenance though. They can get dirty easily and mostly have to be cleaned on a daily basis. If the machines need electrical power, you have to run on power during the period you want to feed. In Photo 26 you can see a self-feeder used in a catfish recirculation system. When the fish touch the iron pin, the feed will fall down accordingly.

The advantage of feeding machines

A feeding machine reduces the amount of manpower at the farm, which is very important in areas that lack manpower or where manpower is very expensive.
If the machine operates well, the fish will eat in a stress-free environment, because there is no disturbance. You can restrict the feeding but not the speed at which they consume!

In Nigeria hardly any farmer uses feeding machines, hand feeding is the more favourable option.

Photo 25. A belt feeder at a fish farm in Abuja. The feed is put on a belt which gradually is rolled up so that the feed is released into the water. The mechanism consists of a clock with a strong spring which runs for 12 or 24 hours.

Photo 26. A self-feeder. This is a feeding machine that is able to adjust the amount of feed to the feeding behaviour of the fish. At present the feed has been finished and the buckets are empty.

Different feed for different growth stages and feed quality

The feed range for catfish shows that there are a lot of different sizes; from 0.1 mm to 9 mm, from fry to brood stock. Appendix 6 shows you the details for each type of feed and the proportions of protein and fat included. It is obvious that the starter feed for small catfish, requires a higher percentage of protein than the feed for bigger and older fish. Apart from the size, most feed millers offer various qualities of feed in different price ranges. The quality of feed is expressed in the protein and fat percentage of the feed. The higher these percentages, the more expensive the feed will be, but it will usually perform better. The feed mills use terms like 'superior', 'essential' and 'economy feed' and farmers have to make choices about which feed to purchase and when to switch to a cheaper kind.

The fact that protein levels and fat levels are high, does not always mean that the feed will perform better than feed with lower levels of fat and protein. Also, the feed miller's choice of ingredients and the way the feed is formulated can make a big difference. For example, if the protein content is based on vegetable protein, for instance soya bean meal, and less on animal protein, a lower protein level diet with animal protein will definitely perform better. It is all about balancing the diet, so that the amino acid profile corresponds to the nutritional needs of the fish. The easiest way to have a balanced diet, is to include a large proportion of fish meal in the diet. Unfortunately, fish meal has become very expensive. Therefore, feed millers use alternatives such as feather meal, meat meal, bone meal, blood meal, soya bean meal, gluten and single components of amino acid like lysine and methionine to balance the fish feed. Feed needs a highly scientific approach to make a balanced cost-effective diet.

The same thing applies for the proportion of fat included. The best option is to use fish oil, but this is very expensive too; alternatives are vegetable fats such as soya oil, but the profile is not as good as fish oil and catfish like the taste of fish oil very much.

Another important characteristic of feed ingredients is the digestibility. Digestibility means how well the feed can be digested and how well it benefits the fish. For example, the leather of a shoe contains a high level of protein, but the digestibility of the leather is almost zero and will not be of any use to the fish!

If you want to know the quality of a feed that is introduced in the market, the best way forward is to test it on the fish. An old saying on this matter is: THE PROOF OF THE PUDDING IS IN THE EATING! However, feed trials at farm level are difficult to do and will not always give a reliable answer about the quality of the feed.

Trials at farms need very consistent daily feeding, good recording and at least for each type of feed two or three different ponds with identical conditions. The test should be carried out against a control which could be the feed which is currently used at the farm. The stocked fish should be from the same batch for all the trials, so that you are comparing the quality of the feed and not the quality of the stocked fish. The mortality is recorded and the water quality, which should be comparable for all the tanks or ponds, is monitored.

Test cropping by taking a sample of fish from a pond in order to obtain an average weight representing the complete fish population of the pond, is not an option. My experience is that this method will not give accurate results and it disturbs the fish enormously. Harvesting of the fish should be carried out on one day for each pond or at least within a week's period. The fish are counted and the total weight is recorded. The results give you total yield, survival rate, average weight, feed conversion rate and uniformity of the fish. The feedback on the fish from the customer or end-user is also a parameter to consider. For example, if the fish have grown well, but have a lot of fat in their body cavities, the customer will not be happy, since the fat is not consumed. In addition, fat makes the fish weak during transport and sales.

7.6 FEED CONVERSION RATE AND THE ECONOMICS OF FEED

The feed conversion rate is simply the weight of feed (kg) you have given to the fish divided by their weight gain (kg) during a certain period. This terminology is not only used for fish but for all animals which are grown commercially. It is the basic equation to find out the economic value of the production, whether it is production of fish, chicken, pork etcetera.

$$FCR = \frac{\text{weight of feed given (kg)}}{\text{weight gain by fish (kg)}}$$

Example 8. How to calculate the FCR?

Suppose that 1000 juveniles of 9 g each were stocked and fed during a period of 5 months with a total cumulative feed amount of 1150 kg. All the fish were harvested and weighed 1015 kg.

To calculate the FCR
Total feed given: 1150 kg
Total initial weight : 1000 juveniles x 9 g = 9000 g (9 kg)
Total harvest weight: 1015 kg

Please recall that: $$FCR = \frac{\text{weight of feed given}}{\text{weight gain by fish}}$$

$$\text{FCR} = \frac{1150 \text{ kg}}{(1015 - 9)\text{kg}}$$

$$FCR = 1.13$$

Example 9. What is the economic value of the FCR?

Using the FCR from example 8 you can calculate the cost to produce 1 kg of fish.
The assumed feed cost is ₦ 375/kg.
Next, the cost to produce 1 kg of fish based on feed cost only:
1.13 x ₦ 375/kg = ₦ 424/kg

Other costs are incurred for juveniles, energy, labour etcetera. The feed costs are called direct costs; They are related directly to the production. Indirect costs are costs which are not related to the production, such as labour; whether or not you sell, you will still have labour costs. In chapter 12 on economics we will go into this in detail.
So, the feed costs don't consist of the purchasing price only. They are calculated by multiplying the performance of the feed by the purchase price of the feed. Many farmers only look at the initial costs of the feed and make the decision from that point of view only, which is unfortunately completely wrong. A cheap feed can turn out to be rather expensive. I will show this to you in example 10.

Example 10. How does a cheap feed turn out to be expensive?
The feed of a local producer costs ₦ 320/kg. After 1 cycle of production the FCR = 1.45. What is the economic value of this feed compared to the feed mentioned in example 9?
The costs to produce 1 kg of fish based on feed costs only:

$$1.45 \times ₦\ 320/kg = ₦\ 464/kg$$

The difference is ₦ 40/kg. This is the direct negative impact of the cheaper feed. It is actually ₦ 40/kg of production more expensive. But that is not all!

To grow the fish in example 10 to the same size as in example 8, you have to add at least 30 days on the 150 days with the cheap feed. This means that the capacity of your production will reduce 20%. In other words, with the better and more expensive feed, the production per year will be 20% more with the same feed gift. So, the running costs are divided over a higher production and will be less, although the feed is more expensive!
Another negative aspect of the cheaper feed is, that it pollutes the water more which results in the need for a higher flushing rate and more energy costs. In addition, when fish get older because the growth cycle is longer on cheap feed, the percentage of eggs in the female fish increases because the females are maturating, which is not desirable for marketing.

Please note the following saying: I'M NOT RICH ENOUGH TO BUY AND USE INFERIOR FEED.

If you start a farm, you have to find out what your expected FCR will be at the farm. This can be carried out after each harvest, assuming that the total fish sales and the cumulative feed for each batch are recorded. The FCR will not vary that much between different batches. This way you can establish an average FCR which you can use for projections. On a weekly basis you can estimate the growth and the individual weight of the fish in each tank, just by dividing the feed with the FCR.

Formula:

$$Growth = \frac{\text{Feed}}{\text{FCR}}$$

When you start selling, you will have a very good estimate of the biomass in the tank. This makes it possible to get the customers on board who can buy all the fish in the tank at once. You will also be able to make a good prediction about the average size of the fish, which is also important for the buyers to know. In chapter 10 on fish farm management I give examples and formats and how to fill them out.

A feed conversion rate of less than 1?

When students come to our fish farm for an IT-training, I always raise the question whether the FCR can be below 1. That means that less than 1 kg of feed can grow more than 1 kg of fish. Many IT-students believe it is not possible but they are wrong!

Example 11. How to get a FCR below 1 for growing juveniles?

Stocking: 5384 pieces of juveniles at 5.2 g average weight
Total kg: 28
Feeding: 233.7 kg in a period of 45 days

Harvest: 4303 pieces of 72.0 g, 111 pieces 25.2 g (runts) and 16 pieces of 312.5 g (shooters)
Total kg: 317.6 kg

FCR = 233.7/289.6 = 0.81

Survival rate = 4430/5384 x 100% = 80%

The runts should be disposed and not restocked with other fish of the same size, since they will not grow and just occupy space.

How is it possible to have an FCR below 1?

The feed has a dry weight of approximately 90%, the fish have a dry weight of 22%. So, when you are growing fish, inevitably you are growing water. Based on dry weight, you will always have losses and the FCR can never be less than 1. The feed used in this result, is a good quality feed with high protein levels (50%) and high digestibility.

7.7 FEED STORAGE

Feed should be stored in a cool and dry place. It's written on the label. Apart from the specification of the feed, you can also look at the production date, expiry date and the NAFDAC-number. All of these are important and should be checked before buying the feed. The bags should be in a good condition, without holes for instance, and the bags should look clean and without dust. The shelf life of feed can be six months to a maximum of a year, as long as that the feed is stored well: far away from heat and moisture.

Buy the feed from a supplier who has a good turnover and a regular supply from the factory. If you have a big farm, you can negotiate directly with the factory, but then you must be willing to buy a trailer of feed, which means 30 tons at once. When you use a 20-ft or 40-ft container for feed storage, the only adjustment you have to make is to roof it. Without a roof the container will become extremely hot inside, which is not acceptable for fish feed, because this will definitely reduce its shelf life.

If you load your feed on wooden pallets, always clean under them each time you load. Stock the feed in layers of 5 bags and maximise the height of the pile at 20 bags, which will result in 100 bags per pallet with a total weight of 1500 kg. You should buy new feed before the old stock is finished but finish the old feed before starting on the latest consignment. "First in, first out" is the motto! So, never stack the new feed over the older feed in such a way that you cannot access the older feed.

Note. Empty feed bags have value, so don't throw them away. Instead keep them in dozens. The price of a dozen empty feed bags can be up to ₦ 200. A farm producing 1 million kg of fish in a year can gain ₦ 120,000 a month from empty feed bags and reduce a lot of waste!

8

REPRODUCTION OF THE AFRICAN CATFISH

8.1 INTRODUCTION

The reproduction of African catfish is not just about following a recipe to get results! It is not simple. Many farmers struggle to get continuously good results. What I noticed in many hatcheries, is that farms get good results in the beginning, but at a later stage the results become worse and worse. As a result, farmers put pressure on the procedures and, in order to save time, they cut corners. Good procedures which are well followed, are the only way forward to get constant results. In this chapter the topics in the field of African catfish reproduction which you need to pay attention to, are discussed.

8.2 BROOD STOCK SELECTION AND MAINTENANCE

It all starts with acquiring good quality brood stocks. Unfortunately, it is difficult to buy them from other farms. The best way forward is to select catfish from good farmers who sell fish for consumption and/or select them from your own fish which you grow for consumption.

The first selection takes place for the general performance of a group of fish in a particular pond. Have they done fairly well, in terms of little disease problems, low mortality and excellent growth? Then you should start to select the biggest females and males from this group, when the time for harvesting and selling the fish from this pond has arrived. Also look out for fish in good condition without deformities. If the fish weigh between 1 and 2 kg, the females already have their bellies full of eggs. This is considered a bad sign. Females less than nine months old, should not yet be mature. Therefore, you should select female catfish for late maturation. As long as the females are not yet mature, they convert all the feed they consume to somatic growth, which means more meat. This is exactly what a grow-out farmer wants. The moment a female matures, the majority of the feed will be used for producing eggs. Some farmers ask for small brood stock because it is cheap, but I believe fish for brood stock should weigh at least 2.5 kg and more and be at least one year old.

In order to avoid inbreeding, never use the females and males from the same batch of fish. Most likely, this fish population will consist of full siblings, which will inbreed when they are used for brood stock.

When you want to start your own hatchery, start with fixing your brood stock facility before building the actual hatchery. Give your fish time to settle in and grow and wean them in a good condition.

The brood stock most often used by farmers, is called the 'Dutch Clarias'. What does this mean? The Netherlands started farming African catfish as far back as 1982. At that time the University of Wageningen was doing research on African catfish in collaboration with a university in Central African Republic (CAR) in Bangui. For this purpose, African catfish were transported to the Netherlands, not only from CAR, but also from South Africa and Israel. This blend of catfish is called 'The Dutch strain' and has been imported several times to Nigeria. Farmers like this catfish strain for their growth, domestication and the ease by which they can be kept together in high densities. You can still purchase brood stock directly from one hatchery in the Netherlands, but it's a tedious and expensive process. In order to import them, you need to acquire an import license from the Nigerian government. They will check your facility and request you, among other things, to make a quarantine facility before the licence is issued.

How to keep your brood stock?

Brood stock can be kept inside or outside, but should not be exposed to too much sunlight and the water temperature should be preferably between 25 and 27 °C. The density of the number of brood stock per m^2 should be at a maximum of 10 to 15 pieces per m^2. Stocking them at a density of less than 2 per m^2, can cause problems too. Freshly stocked premature brood stock in low densities start fighting, which causes a lot of problems, such as infections and mortality. Fighting cannot always be avoided though. Whenever you buy catfish brood stock, add 2 kg/m^3 of salt to the water in order to avoid infections and improve the healing of the skin after fighting.

The water quality should be perfect, with ammonium levels of less than 5 mg/l and a pH between 6.5 and 8. Females and males should be mixed together in one tank. Catfish produce pheromones which stimulate the development of the gonads in both sexes. Stock the brood stock in several ponds instead of putting them all in one; don't put all the eggs in one basket!

Also, it is wise to grow a group of premature catfish to maturity for the next generation of brood stock at the same time. When the females are used, they should be put in a different tank to recover and grow another set of eggs. Usually after three to four months, you can select ripened females again for another spawning.

A question many farmers ask, is: "How much brood stock do I need to get a certain quantity of production?" The answer depends on so many factors, but the most important one is the success rate of the spawning. Farms which perform very badly, can use up to 200 eggs to grow 1 catfish juvenile. Others are more efficient and use only 5 eggs to grow 1 juvenile. The fecundity of catfish females is very high compared to species like Tilapia. The body weight of a female consists of 5 to 15% egg mass and 1 g of eggs means up to 500 pieces.

For example, a 4 kg female contains between 200 and 600 g of eggs, which is equivalent to 100,000 to 300,000 pieces. If the spawning and rearing procedures are carried out well, this results in 20,000 to 60,000 juveniles. The female can be used at least two times a year. So, one good female can yield up to 100,000 juveniles a year!

Male catfish usually are sacrificed in order to collect their sperm, but an operation is also possible. In this case you need aesthesia to put the fish to sleep and a sharp surgical knife to make a small incision in the belly. Then you take out part of the testes by cutting a piece off or suck it out with a syringe, if the testes are very big and full. The advantage of operating is that you can spare the males and use them again. Also, it's good to save the males when you want to carry out a selection programme to improve your catfish strain.

The sperm of the male varies a lot in quality and it cannot be accessed otherwise than opening the belly. Good testes look milky and contain a lot of fluid, which looks milky too. Good testes can contain up to 10 ml of sperm and 0.5 ml of sperm can fertilise at least 100 g of eggs. The concentration of good sperm can be up to millions of sperm cells per ml. Males, kept for three years or more, will have a reduced sperm quality. The testes become spongy and the sperm becomes very thick and less fertile. Females can be kept much longer, if taken good care of. When females become older and bigger, they will give more and bigger eggs and the quality usually improves, especially when they are used for stripping the eggs during this period.
If you are serious about starting a brood stock programme and keep good records of brood stocks and their performances, you cannot do without a microchip tagging system (Photos 27 and 28) to identify your males and females.

To improve a strain by keeping a large number of brood stocks in your ponds, selecting them on performances by following the offspring in a professional way, is almost impossible at farm level and will cost a fortune. To get results from improving brood stock and offspring, will take a couple of years and, improvements of 5% are great achievements for a researcher, but not really for the farmer. Small differences at farm level will not be spotted without serious record keeping. Improvement of brood stocks should be an assignment for institutes or big feed companies.

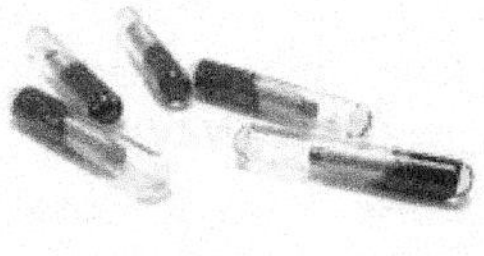

Photo 27. Microchips which are injected into the catfish.

Photo 28. A microchip reader reads the microchip inside the body of a gravid female catfish. The history of this female is written down in a book and provides information on the success of former spawning results.

Feeding the brood stock

Some years ago, not many farmers used brood stock feed but gave their brood stocks ordinary grow-out feed and sometimes added some crayfish or fish to the diet. Nowadays, it is necessary to use brood stock feed because the grow-out feed doesn't contain the required ingredients for brood stock anymore. The proportions of fish meal and fish oil have been reduced or are completely absent because these ingredients have become so expensive, and therefore have been replaced by alternatives, such as animal meat meal, hydrolysed feather meal etcetera.

Brood stock feed is characterised by the inclusion of the best fish meal, fish oil and micronutrients. Although not tested on catfish, but on species like salmon and trout of the Salmonids family, these ingredients increase the fecundity in both sexes, the hatching rate and the health status of fry (see Appendix 1 for catfish growth stages). In addition, it decreases deformities in fry. Don't minimize the quality of feed for your brood stock, it is your starting point to get good quality fry. When using brood stock feed, the colour of the eggs becomes dark-green to brownish, which means that they are good and viable.
Feeding should be carried out at least two times a day to satiation level, which is normally about 0.5% of their bodyweight. When you want to select brood stock, stop feeding 24 hours before you start.

8.3 WATER: QUALITY, STORAGE AND TEMPERATURE

Water quality

Catfish hatcheries should use either borehole water, well water or city council water. The water quality is considered to be very important for successful spawning and good survival rates. I do not have a complete list of parameters which are considered to be important, but some can have a really negative impact on fry and fertilised eggs.

These parameters are:

- Levels of iron above 0.2 mg/l
- A German hardness level above 7, each unit of German hardness is 17.8 mg/l $CaCO_3$
- pH above 8
- Chlorine gas above 0.05 mg/l, when city council water is used
- High levels of bacteria from a contaminated borehole and/or a contaminated reservoir

If the water has been tested, and the laboratory advises that it is useable as drinking water, it will also be suitable for the hatchery, assuming that the pH and the hardness are below the levels given above. So, when you start a hatchery, the first thing to do is to check the availability of good quality borehole water. Actually, before investing in anything else this is the first thing you should do. Nigeria has laboratories which can check this for you. Apart from testing the water, never use anything other than PVC or PE-pipes for water supply. Galvanised or copper pipes bleach out chemicals such as cadmium, zinc and copper which are certainly toxic for eggs and fry.

Water storage

It is advisable to use the borehole water indirectly via an overhead tank. Plastic tanks made from polyethylene (PE) are commonly used to store water and they are durable if well supported underneath.

Borehole water contains a lot of carbon dioxide but usually has little or no oxygen. Using this water directly for the eggs or fry can result in problems. The best way is to oxygenate the water and eliminate the carbon dioxide before it enters the storage tank. This can be carried out by a small degassing tower on top of the reservoir. The storage tank should be cleaned every two to six months; accumulation of debris on the bottom of the tank should be avoided and removed.

Water temperature

The water temperature is very important for good hatching results. It reflects the air temperature and fluctuates per season and region. In the tables below, you can see the average day and night temperatures for Abuja, Jos and Lagos (Figure 22).

Average minimum and maximum temperatures over the year in Abuja.

Abuja - Average temperatures												
Month	Jan	Feb	Mar	Apr	May	Jun	Jul	Aug	Sep	Oct	Nov	Dec
Min (°C)	20	26	24	25	20	18	22	18	18	21	16	16
Max (°C)	35	37	37	36	33	31	29	29	30	32	34	35
Min (°F)	68	79	75	77	68	64	72	64	64	70	61	61
Max (°F)	95	99	99	97	91	88	84	84	86	90	93	

Average minimum and maximum temperatures over the year in Jos.

Jos - Average temperatures												
Month	Jan	Feb	Mar	Apr	May	Jun	Jul	Aug	Sep	Oct	Nov	Dec
Min (°C)	14	16	18	19	18	18	17	17	17	17	16	14
Max (°C)	28	30	32	31	29	27	25	24	27	29	29	28
Min (°F)	57	61	64	66	64	64	63	63	63	63	61	57
Max (°F)	82	86	90	88	84	81	77	75	81	84	84	

Average minimum and maximum temperatures over the year in Lagos.

Lagos - Average temperatures												
Month	Jan	Feb	Mar	Apr	May	Jun	Jul	Aug	Sep	Oct	Nov	Dec
Min (°C)	22	24	24	24	23	23	22	22	22	22	23	22
Max (°C)	32	33	33	32	31	29	28	28	29	30	31	32
Min (°F)	72	75	75	75	73	73	72	72	72	72	73	72
Max (°F)	90	91	91	90	88	84	82	82	84	86	88	

Figure 22. The average temperatures of Abuja, Jos and Lagos (source: World Climate Guide).

The water temperature is considered to be optimal for hatching and rearing fry when it is 29 to 31 °C. The periods in December, January and February are considered too cold for successful spawning, if measures to heat the water during the night are not taken, especially for Abuja and Jos, because the harmattan in these regions has a serious impact on the temperature from evening till dawn. Lagos has the best climate for a catfish hatchery. The nights are not too cold and the fluctuations in temperature are less than in the middle belt or the highlands, such as Jos. When the water temperature is 25 °C or below, the hatching of eggs and even the absorption of the yolk sacs will be significantly delayed. Also, the fry start eating only after 2.5 to 3 days instead of after 2 days. Their behaviour will be sluggish and the fry will not be as strong as when the water temperature is 30 °C.

The height of a hatchery building should be less than 4 m and it should not be furnished with a ceiling. The roof should be dark coloured so that it will absorb heat and the water will at least be heated during the daytime. Minimise ventilation, especially during the night. The building should be air tight. During harmattan the humidity is very low. This causes a lot of evaporation and as a result the water cools down even more. When the building is airtight, you increase the level of humidity which will result in less evaporation and the water will cool down less.

If maintaining the temperature of the water above 27 °C is a serious problem for you, you have the option to add extra heating. Glowing charcoal can be used to heat the air. Be sure the charcoal has stopped fuming before putting it in the hatchery room. Heating the water with an electric heater is another option to increase the water temperature, but make sure that the heater cannot be in direct contact with the fish and that it does not touch any walls. If you use electrical heaters in the water, always turn them off before you put your hands in the water. Some heaters have a built-in thermostat, others are just heating the water till you turn them off.
Solar heaters are another option for heating the water, and as a matter of fact they work very well, because Nigeria has a high level of sunshine. If you make one yourself, the biggest challenge you will face is deciding which material to use for exchanging the heat. If you choose PE or U-PVC and the water is not going through the exchanger, the material will melt and the pipes will start leaking. Metal exchangers, for instance those made of copper, cannot be used since they poison the water. The only pipe material I believe is suitable, is stainless steel. Stainless steel is non-toxic, transfers heat very well and is resistant to high temperatures. The principle is simple: wooden casing, an absorber, an exchanger and a transparent plate of glass or Perspex are all that is needed.

8.4 SPAWNING PROCEDURES

Natural spawning

Before explaining the artificial procedure for spawning catfish used at fish farms, it is good to know how African catfish spawn in their natural habitat. The African catfish has a spawning cycle and produces offspring only during a couple of months each year. The period is initiated by the start of the rainy season when rivers, dams and lakes are flooded. For Nigeria, it starts in March, at the time when the water temperature reaches its highest levels of the year. The female and male catfish will have their highest GSI (gonadosomatic index). This means that during this time, the weight of the testes and ovary occupies the largest percentage of their bodyweight, which is 1% for the males and 10% for the females. High water temperature is also beneficial for the growth of fry and for the increase

of their natural food supply. During the rainy season, the flooded areas contain increased levels of organic material, which enhances the growth of zooplankton, the natural food for the freshly hatched catfish fry.
Catfish mating usually starts in the evening or in the early morning and takes place on the shoreline of a lake, river or dam. Both sexes are close to each other, vigorously swimming and releasing the sperm and eggs in abundance. The sticky fertilised eggs stick to the bottom and to underwater plants. Catfish don't have parental care, so the hatched fry are on their own from day one.

In captivity catfish rarely spawn because the last step of egg maturation and ovulation do not happen, so the eggs are not released. Although, on rare occasions spawning does happen. It is initiated when gravid brood stock is transferred from an environment with dirty water to one with cleaner water. It's annoying when this happens, because if the females do release their eggs, you cannot use them for a couple of months, and the eggs released in the fish pond cannot be collected. The change of water quality actually imitates a rainstorm; the reduced level of dissolved chemicals boosts the brooders.

Artificial reproduction of catfish is carried out by injecting a hormone into the female. It is a standard procedure and it is temperature related. The hormone is a gonadotropin releasing hormone and can be a pituitary gland or a synthetic product such as Ovapitine, Ovaline or Ovaprim. This hormone induces the final step of ripening of the eggs, called vitellogenesis. In this final step the eggs are provided with a yolk and absorb water.
The time between injecting this hormone and stripping the female, takes around 10 hours, but can be less if the water temperature is higher than 26 °C, and more if the temperature is less than 25 °C.
The best time of stripping has been found by trial and error. As a farmer you will get to know the best time for stripping from your experience of previous spawnings. The release of the first eggs in the tank, where you have isolated the female brood stock, is a good sign. If you try to strip the eggs out of the belly too early, it will be difficult. The eggs look somewhat dry and you will have a pile of eggs in the egg collection tray instead of evenly spread layers of eggs.

Step by step procedure for artificial reproduction

Preparing the system/tank: disinfection of the incubator

In order to get continuously good results with catfish hatching, it is important to clean and disinfect the system in which you put your fertilised eggs. This system is called 'the incubator'. Most farmers use the flow-through system for incubating the eggs, but some use small recirculation systems. The latter are more difficult to disinfect, in particular the filtration unit since more components are involved, and the disinfection procedure is different.

Flow-through system

Most farmers use white plastic tanks to incubate the eggs and get good results. However, after each spawning and transferring the fry, you immediately have to clean the tanks thoroughly with a brush and apply either calcium hypochlorite, which is also used in swimming pools, or hydrogen peroxide as a disinfectant. The first disinfectant comes as a white powder while the latter is a liquid of 35% industrial hydrogen peroxide.
Apply this to all the surfaces of the tank in a 1% solution of calcium hypochlorite or a 3% solution of hydrogen peroxide. This can be carried out with a rag or by spraying it with a knapsack sprayer. Allow the disinfectant to remain on the surface for at least one hour. Follow the same procedure for the hatching nets or hatching trays, the filter and the PVC-materials. Use the same rag or brush to dispose of the disinfectant with abundant clean water. Fill the tank with water 12 hours before putting in the fertilised eggs.

Recirculation system

Hatcheries using recirculation technology also have to be disinfected completely after each spawning. So, after the fry are transferred, all the components of the recirculation system, such as the sedimentation tank and the biofilter unit have to be thoroughly cleaned with a brush and/or a high-pressure cleaner. The best thing to do is to take small components outside and clean them with a high-pressure cleaner while the tank and sedimentation tank are cleaned inside with a brush. After all the components of the system have been cleaned, put them back and fill the tank completely with water and allow it to circulate. Add the disinfectant, either calcium hypochlorite at 30 g/m^3 or 1 l/m^3 of a 35% hydrogen peroxide solution for 12 hours. Flush, rinse and fill the system again, and allow the water to circulate for another 12 hours before putting in the water for starting up another spawning.

Notes.

- Calcium hypochlorite and hydrogen peroxide are both strongly oxidizing chemicals, which attack any organic matter including all living organism. The substances change the composition of proteins and enzymes in living cells which will result in dysfunction and death.
- Always use personal protection if you weigh out and apply these chemicals, because they are also harmful for human beings!
- If your water is very soft (German hardness of less than 4), you are lucky, because systems and tanks filled with soft water will not get dirty rapidly and are very easy to clean. As a matter of fact, spawning in soft water is more successful and the growth of fungus on dead white eggs occurs less frequently.

Selection of females and males prior to injection

You have drained a brood stock tank to select female and male catfish for reproduction. You can identify a gravid female that is ready for injection by her swollen red papilla. It should be possible to strip out some eggs with little force and you will see that the eggs have a migrated nucleus, which means that the core of the cell is situated at the periphery of the egg. The eggs have a diameter of approximately 1 mm. I prefer to select the female catfish in the early afternoon and inject them with the hormone in the evening. In the same way, the male catfish, which should be at least 300 days old, are selected. The maturation of the male is difficult to discover from the outside. Only dissection will show you if the testes are well developed with milky milt. Each female and male catfish should be put into separate tanks located in a quiet environment. The water temperature should be maintained at around 26 °C, both day and night. A very suitable isolation tank is the white rubber 1000-l tank, used to transport chemicals and sold afterwards in many places in Nigeria. Allow the fish to have at least 300 l and again don't stress the brooders.

Injection of the female

Just recently, I noticed that in some farms the female fish get infected after injection. The location where the needle is inserted, swells up and results in a wound. The fish can easily die from it, which is a big loss for a hatchery, especially if the female has been giving you good spawning results. When I was doing the hatchery and the spawning operations myself, I never experienced this infection and I wondered why! So, I asked the hatchery manager, whether he was using new needles with each spawning. He told me: "Yes the needle is still sharp!" So, I asked him: "If a doctor uses a needle, which he has just recently used, with the comment 'the needle is still sharp', you will buy that idea?" He said: "No way, I don't want to die of HIV or any other disease." Please, use a new needle every time for each spawning. The cost of a new needle is just ₦ 20, don't be penny wise and pound foolish!

At around 8.00 pm inject the hormone into the female catfish; the dosage is 0.5 ml for each kg of fish. Use a 5-ml syringe and a needle with the specification 23Gx$^{11/4''}$. Needle and syringe are both readily available at a good pharmacy. The hormone can be bought from companies which sell fish feed, drugs etcetera.

Never insert the needle in the middle of the fish towards the central spine. It can start bleeding heavily, because the central vein is located in this area. Insert the needle at an angle of 45 degrees away from the head in the thickest part of the muscle. The hormone is viscous and the syringe should be pressed gradually in order to empty it inside the muscles of the fish. After gently withdrawing the needle, you massage the spot where the needle entered for some seconds, before putting the fish back in the isolation tank. This procedure is easy to carry out without using force on the female. She remains calm as long as you handle her gently.

Males are not injected, but kept in the isolation tank ready to use for the next morning. After injection, the females become a bit restless and can start to jump. Ensure that the fish do not jump out of their holding tanks to keep them safe. The female fish often change colour and gain weight, because of the water absorption during the egg ripening process.

Collection of the eggs

In the morning, the female fish will be ready for stripping. Take the female out, but if anaesthesia is available, you should first tranquilize her. MSS-222 (tricaine methanesulfonate) is an anaesthetic for fish, including catfish. It is added to the tank at a ratio of 2 g/10 l. After some minutes the female fish will stop breathing and turn on her side. Take the fish out, gently supporting the entire body, and wash off the water containing the anaesthetic, with clean water. Place the fish on a smooth table and cover it with a clean wet towel; the towel should not leak any water but also should not dry the skin of the fish, because this will damage its slime. The stripping can now proceed. The female is stripped until the eggs are not flowing out anymore. When blood is coming out with the eggs, it is a sign the ovaries are getting empty. Afterwards the female is put back into the tank she was isolated from, but not before its water has been refreshed. Then the eggs are weighed.

Collect the eggs on an egg collecting tray, preferably made from glass or ceramic plate. A smooth surface does not allow the eggs to stick to the tray surface easily. If you don't have anaesthesia, you have to apply a bit force to stop the female from moving her tail and wasting or dirtying the eggs on the collecting tray. Most of the time the female will remain relatively calm. The eggs should be stripped as cleanly as possible; without slime, water or faeces (Photo 29).

In the afternoon the female must be taken out from her tank again to strip the remainder of the eggs which should be discharged. After stripping the ovary should be completely empty, otherwise the eggs left behind, will start to rot which causes serious problems for the fish.

Photo 29. Stripping the eggs from the female catfish.

Collection of milt (sperm)

There are two ways to collect the sperm from the male; either by sacrificing the male via dissecting the testes out of the belly, or by keeping the male alive, and collect the sperm by taking a part of the testes out or sucking the sperm out of the testes with a syringe during surgery.

Sacrificing the male is the most usual method and it is relatively simple. The best and easiest way to kill the male is by hammering him on the head. It seems cruel, but if you do it with force, the fish will die almost instantly. Clean the belly with a towel and make an incision from the anus upwards with a clean surgical blade. Open the cavity and look for the testes which are located deep inside the cavity, close to the spinal cord. Use the surgical blade to cut the testes out of the tissue which is holding them in place in such a way that you keep the testes complete. There should be no leakage of sperm from the testes. Collect the testes in a small dry cup and cover it with tissue paper. Then put it aside in a safe place where it cannot fall or come into contact with water.

Note. Sperm of fish is inactive as long as it has not been in contact with water. The moment the sperm cells in the sperm come into contact with water, the energy locked inside activates them to start surging for eggs. This will last for only about 60 seconds. Afterwards the sperm is not fertile anymore! Also, eggs which have been in contact with water for a few minutes, are not fertilizable, because the micropyle, the opening where the sperm has to enter the egg, will have closed. Some fish species have a special attractant on the chorion, the outer membrane of the egg, which directs the sperm cells into the micropyle of the egg. So, it is important to collect eggs and sperm in a clean cup or bowl without any contact with water!

Note. Dilution of sperm in a physiological salt solution of 0.9% sodium chloride in water to enhance mixing of sperm and eggs before adding water to trigger the fertilisation, is also possible. Be sure the solution is fresh and has a temperature of 28 °C.

More about milt

Sperm or milt can be preserved in a fridge at 4 °C for one day, but it's advisable to use the milt directly after it has been taken out of the male. Sperm can even be frozen and remain viable for many years. This process is called cryopreservation. This technique consists of using special chemical solutions which are added to the sperm, and liquid nitrogen to keep the milt as cold as -173 °C in a special container. For Nigeria this technique is not feasible at farm level, but in aquatic research institutes it could be interesting to use this method for improving the genetics of the catfish.

Sperm quality can be tested under a microscope; put a small drop of sperm on a glass slide and look at it with the 100-x magnification. If there are plenty of sperm cells in the milt, you will see millions of little grey balls which you can activate with a little drop of water. The sperm cells should swim actively for some seconds. This is the sign that the sperm is fertile and is able to fertilise the eggs.

Fertilisation process

The eggs are divided between different bowls with no more than 200 g in each bowl. Fertilise 200 g of eggs at a time with at least 7 to 8 drops good quality white milky sperm. More sperm will not increase the fertilisation rate, but it will not cause any harm. Make small incisions in the testes and by a little squeezing, the sperm will come out and must be directly put onto the eggs. Without water nothing will happen, but immediately after the sperm is put onto the dry eggs, you have to activate the sperm by putting water in the bowl. For 200 g of eggs half a litre of water is enough. Stir the solution gently for about three minutes, so that the sperm can be optimally mixed with the eggs and fertilise them. The water you use to activate the sperm, should come from the tank which you use to put the fertilised eggs in.

Afterwards, spread the eggs over a mosquito net. Most eggs will stick to the net, but some will go through it and stick to the bottom of the tank. 200 g of fertilised eggs need a surface of 0.25 m^2 to be spread onto. It should be a single layer of eggs on the net, so that the bad unfertilised eggs on which fungus will start growing, do not infect the good fertilised ones.

Incubation time till hatching

The incubation time till hatching is about 24 hours, but it can vary a lot, depending on the water temperature. The best temperature for incubation and hatching is 30 °C. The acceptable range lies between 27 and 31 °C. Below 25 °C the hatching will take too long and will give poor results. A low temperature makes the fry weak and less active. They will stay where they have been hatched and will not swim to the corners of the tank. They will just drop onto the bottom and get infected with fungus or bacteria.

After the fertilisation procedure, it takes about 10 to 12 hours before you can check if the eggs are actually fertilised. The nucleus grows into a round-shaped line at the periphery of the yolk sac, forming the tail of the embryo. At that moment you can distinguish the eggs which have not and which have been fertilised. The unfertilized eggs become white while the fertilised ones are still greenish, shining and the development of the embryo is visible.

Usually, the hatching rate differs from the fertilisation rate. If the eggs are strong, the hatching and fertilisation rate will be close, but normally the hatching rate will be not more than 50% of the fertilization rate. When most of the eggs have hatched and dropped out of the mosquito nets, you must remove the nets from the water. If you wait too long, the water will get whitish and fungus will develop, which has a negative impact on the yolk sac of the fry.

Clean and rinse the nets straight away!

Note. Don't mix the eggs of different females in one bowl or in one incubation tank and use one testis for each bowl of eggs. The offspring from the hatching should be full siblings, so that the genetic variation in the fry is as low as possible and the growth will be as uniform as possible.

One day I visited a farm in Mali. They have a well-equipped catfish hatchery and we prepared for a spawning of Hetero-clarias: selecting the brood stock, preparing the incubator by cleaning, disinfecting and rinsing, and checking all the materials needed for the actual spawning, such as anaesthesia, bowls, hatching trays, syringe, surgical blade and hormones. Another check was carried out on the availability of starter feed, salt and Artemia, also called 'brine shrimps', which are a good starter feed for fry. Well, in the evening we wanted to inject two females; one Heterobranchus and one Clarias female. We checked in the fridge before and knew that we had at least three bottles of Ovaprim still in their packaging, so we thought there was no problem. How surprised we were, when we found out later that all the bottles of Ovaprim were used and almost empty! So, we had to postpone the hatching procedure. The next day we started looking for the bottles with the hormone!

I notice in many farms that empty packages are never disposed of but put back into the fridge, cupboard or any place where you usually can find them. This provides a false sense of security and it's annoying too. I asked the farm manager why all those empty packages of hormones, artemia, starter feed, antibiotic etcetera are never disposed in the dustbin? The answer was clear: if the boss asks about these materials, the staff can show that they have used them by showing him the empty packages. I suggested having a separate cupboard for this, so that you can at least keep a clean record of your real stocks!

When you prepare things for spawning, do it well; make sure all the materials are available and put in the right place before you start!

What may cause unsuccessful hatching?

In most cases unsuccessful hatching has multiple causes from a variety of angles and therefore not easy to spot and to correct, but I try to list the issues which usually go wrong:

- Brood stocks are of inferior quality because they have not been given good feed and/or have been underfed and/or are less than 12 months old.
- The female does not release her eggs. There are multiple reasons for this: the incubation temperature of the water of the holding tank is too low (less than 24 °C), wrong dosage or expired hormones, badly handled and stressed brood stock or the female is just not yet mature.
- The collection of stripped eggs contains a lot of white eggs; this is a sign of regression of the eggs or off-season spawning. In nature catfish only spawn during the rainy season, so the dry season is not the period catfish have their best eggs!
- The eggs look good but turn white within eight hours. Most likely the disinfection of the tank has been carried out but it has not been rinsed well afterwards; traces of chlorine can kill the eggs within a short period of time.
- The eggs delay hatching if the temperature of the water is 25 °C or below. The incubation time between fertilisation and hatching can become up to 30 hours or more. The yolk sac-fry will be sluggish and are not likely to swim to the corner of the tank. They stay at the spot on the bottom of the tank where they dropped down. Then, the fry can get easily infected with bacteria or fungus. This will reduce the initial survival rates and it will also cause reduction of the number of fry during the first two weeks of growing.
- The water quality also plays an important part in successful hatching; preferably the pH is around 7, the German hardness not higher than 7 and the iron content is less than 0.2 mg/l. Spawning of catfish eggs proceeds best in soft water. Another problem with hard water is the much more rapid and severe growth of fungus which infects a lot of hatchlings
- The storage tank is very dirty and causes a lot of bacterial influx into the system. The storage tank of a hatchery should be cleaned at least every three months. When you have hard water, containing a high level of calcium and magnesium, the storage tank gets dirty very quickly. The limescale (calcium) deposits on the walls of the tank and pipes are a breeding ground for bacteria and will have a very negative effect on results for freshly hatched eggs and fry.
- Oxygen levels are too low because water is not flowing through the tank constantly. If you use the flow-through system, it's advisable to have an aerator in the tank to support the oxygen level in the incubator and during the growth stage of the fry.

Yolk sac absorption till first feeding

The yolk sac-fry are not yet developed into a fish. The mouth has not been formed and the digestive system has not been completely developed. Over the course of two days at a temperature of 28 to 30 °C, they absorb their yolk sac and change into swimming fry able to eat. In this period the fry avoid light and move away from it, which is called 'negative phototaxis'. In the tank you will see them clustering in dark corners. It is important that the tank for the fry has no drain where they can hide in. The fry will certainly die here, since the oxygen level will be too low for them. In this period, the fry also start to show pigmentation; they turn from greenish to brownish. When the fry swim and cluster into corners, it's a good opportunity to clean the bottom of dead eggs, fungal and bacterial growth etcetera. This can be done by sucking this waste out with a small 4 - 5 mm hose. After two days the fry are ready to start eating; they come out of the corners and swim up the water column, searching for food. They will be swimming everywhere, not in schools just erratically; that's why they are called 'swim-up-fry'.

8.5 FEEDING FRY

First feeding

In the olden days "What to feed the swim up-fry?" was one of the most crucial questions. Feed from the factory was rarely available and starter feed was not even developed at that time. So, farmers breeding catfish used natural plankton, egg yolks or egg whites and ground fish meal as the first feed, with mixed results. In the '80s Artemia salina cysts became available and they appeared to be a very good starter feed for catfish. Till today you can use Artemia salina to start feeding the fry at their initial stage, but this is not common in Nigeria because of the low availability and high price (Photos 30 and 31). In Appendix 8 you will find the recipe for making starter feed from Artemia cysts. An alternative is encapsulated Artemia cysts, which are available in shops specializing in fish feed products.

At present several feed factories produce or import good starter feed. They are expensive (₦ 9000/kg in 2018), but when you calculate it, it is preferable to use good starter feed instead of inferior products, as long as you get improved survival rates and good healthy growth. Appendix 6 shows a list of starter feeds with specifications.

Live Artemia have a big advantage over dry starter feed because they are very much appreciated by the catfish fry and they stay alive in the water for at least 30 minutes. Live Artemia don't pollute the water like dry feed does. Overfeeding of Artemia is not a big problem, whereas overfeeding dry feed causes excessive bacterial growth, which means trouble for the catfish fry. The feeding frequency of catfish fry should therefore be every three to four hours and in small quantities at the time.

You can see their stomachs getting bigger after feeding and if you give them orange feed, their bellies will turn orange too. When the fry are satiated, they become less active in the water column and rest at the bottom of the tank. After some hours, they are normally already in search of feed again and swim up. So, if they do not come up after some hours, there is a problem. The water temperature plays an important part in the response and growth of the fry. Preferably, the water temperature should be 30 °C all the time.

Photo 30. Hatching Artemia, to make starter feed for catfish swim up-fry.

Photo 31. Harvesting Artemia, starter feed for catfish swim up-fry.

The economics of the use of so-called 'expensive' starter feeds

Many farmers don't want to spend money on the best starter feed; they believe they can never make a profit if starter feed is expensive, and therefore use alternatives. But it is important to achieve good growth and good survival. In that case, an expensive feed will pay for itself and it will actually become a cheap feed!

Example 12. What is the feed cost to produce a fingerling of 0.6 g?
Suppose the starter feed for the first week is ₦ 14,000 per kg (particle size 0.1 mm) and the second week you use a starter feed of ₦ 11,000/kg (particle size 0.2 mm). After these two sizes you can use an ordinary starter feed for catfish as long as the size is between 0.2 - 0.3 mm. This feed costs about ₦ 1000/kg.

FCR = 0.6 which means 0.6 kg of feed is needed to produce 1 kg of fingerling.
0.1 mm is used for fry from 2 mg to 25 mg in week 1 (growth 23 mg).
0.2 mm is used for fry from 25 mg to 80 mg in week 2 (growth 55 mg).
0.2 - 0.3 mm and 0.5 mm is used for fry from 80 mg to 600 mg in week in week 3, 4 and 5 (growth 520 mg).

Feed needed for 1 fish of 0.6 g:
0.1 mm: 23 mg x 0.6 = 14 mg of feed
0.2 mm: 55 mg x 0.6 = 33 mg of feed
0.2 - 0.3 mm: 520 mg x 0.6 = 312 mg of feed

Cost of feed per size for 1 fingerling:
0.1 mm: 0.014 x ₦ 14,0/g = ₦ 0.2/piece
0.2 mm: 0.033 x ₦ 11,0/g = ₦ 0.36/piece
0.2 - 0.3 mm: 0.312 x ₦ 1,0/g = ₦ 0.312/piece
Total cost for 1 fish of 0.6 g = ₦ 1.0/piece

Not all fry will make it to fingerling and therefore you should multiply ₦ 1.0 with 2. The cost to produce 1 fingerling will still be ₦ 2,0 or less a piece!

8.6 GRADING ACCORDING TO SIZE

Catfish are carnivorous and cannibalism is normal, if the situation is favorable for it. To avoid cannibalism in the early growth stages, you need to grade the fish according to size. If the fish are of the same size, cannibalism will be reduced. Cannibalism can already start from the first day the fry are eating. It will be especially enhanced in less optimal situations, such as having weak fry or dead fry in the tank. The fry will not all develop at the same speed. Some fry will still be eating their yolk sac while others are already looking for food. If the first feed is too early, you already create a difference in size from the start! Normally when the fry are healthy and feeding is optimal, not too less, not too much, cannibalism will not be a big problem, since they all grow at the same pace. That's why I always recommend growing batches of full siblings and not a mixture of many different genetics, which enhances different growth rates.

During the first two months till the juvenile stage, the fish have to be graded at least two to three times. Grading can be carried out in different ways. You can decide to take out the shooters only and leave the rest of the batch to grow on. Or, you take out the shooters and grade the remaining fish into two different sizes. Shooters are at least three times bigger than the average size. They really eat other fish and sometimes you can see those shooters vomiting complete catfish when you start grading. They take a bite from the head and swallow their prey whole. A shooter of 12 g can eat a fish of 4 g! If you see many head injuries in a group of fish, it's an indication that there are many shooters inside. Shooters usually don't eat dry feed, but start hunting any time they are hungry. They even fight each other and can die as a result.
I don't consider shooters good brood stock, although they grow much more quickly than the others. But this is not a result of genetics but of the excellent food they get. If shooters are not taken out in time, cannibalism will reduce your fish population up to 50% in two weeks' time!
Grading can be carried out in different ways. Apart from a grading table you can also use an adjustable grader or net grader (Photos 32 and 33).

Photo 32. A table grader for post-juveniles.

Photo 33. An adjustable grader and a net grader for fry, fingerlings and juveniles.

8.7 SELLING FINGERLINGS AND JUVENILES

Health status of the fish

The fingerlings and juveniles you sell your customers, should be in a perfect condition and well fed. Some hatcheries stop feeding their fish well, as soon as they receive a booking and get paid a deposit or full payment. After all, the fish have already been sold. The customer has just to collect them and the fish are already at their selling size. If this period takes time, the fingerlings or juveniles become weak and lean; actually, what you have been taking care of for two months, you spoil in a matter of days, just because you don't want to buy another bag of starter feed. Discuss with your customer that the price of the fish depends on their size. If the fish are not collected by the customer in time, which happens a lot, and the fish become bigger, the price will be adjusted. If the fish have been sick, be sure that the sickness is completely over before you sell, because otherwise the transport and stress of a new environment can trigger the sickness to return. Your customer doesn't know how to handle this problem and will certainly encounter severe losses; another customer who will not buy from you again!

Grading

If you have a hatchery and you sell your product to customers, then it's an obligation to grade the fish very well. Most grow-out farmers don't have the means to grade and stock fingerlings

and juveniles in a big pond with the projection to keep them in this pond till harvest!
In any group of fish, you will have runts: fish which don't grow at all. When the fish have for example an average weight of 10 g, the runts are only 2 g. In Nigeria farmers are often reluctant to throw these little fish away and still keep them or worse, stock them with other fish from a different batch. Fish which don't grow, are a liability. You have to get rid of them as quickly as possible. Don't waste feed on them and don't sell them to spoil your name!

Counting or weighing method

Counting the fish one by one is an accurate way to collect the number of fish a customer wants to buy, but when it takes too long or if the number of fish is over 10,000 pieces, it is not the right method. Most farmers believe counting the fish is the best and most accurate way, but by counting the fish many things can still go wrong. Usually they use the grading table for counting. The table is smooth and the fish are put on top of it. They count till one hundred and put a stone aside; later on, they count the stones to estimate the overall number of counted fish.

I personally prefer the average weight method. Because the shooters should have been taken out, you take the average weight of the fish after grading from each tank. At least 300 pieces are counted and weighed on a scale with an accuracy of 1 g. The average weight will be the total weight divided by the number of fish. If you know the average weight, then you divide the total weight of the fish by the average weight to obtain the number of fish you want to sell. This method doesn't stress the fish as much as counting them and it's done quickly. Automatic fish counters can be used too but are very expensive.

Example 13. How to collect 5000 pieces by the average weight method?
You weighed 315 fish and you got a total of 500 g.
The average weight is 500/315 = 1.59 g.
Now, you want to sell 5000 pieces, so the total weight you want to sell is: 5000 x 1.59 g = 7,950 g or 8 kg. So, you weigh 8 kg of fish for your customer. Usually you give the customer 2 to 3% extra to be sure he gets the number he wants.
In this case you weigh out for the customer an extra of 0.16 to 0.24 kg.

The problem with the average weight method is the water attached to the fish. The smaller the fish to be sold, the bigger this problem will get, because the amount of water attached to the fish, provides false information about the average weight and the total number of fish. But, the smaller the fish, the more damage you will face when using the one by one counting method.

Packaging and transport

Some years ago, I was invited to evaluate a fish farm in Ogun State close to Ijebu Ode. As I was already staying in Abeokuta, it was a convenient journey, because the general manager of the fish farm picked me up from there. On our way to the farm, he explained about his plans. He wanted to upgrade his hatchery, because he faced some challenges buying the fingerlings and just recently had lost a lot of fish due to transportation. I asked him what happened and he told me the story. The fish he always buys originate from a farmer friend and their quality is good. Since the trip between the farms is short, they quickly weigh the fish and put them in those 50-l plastic bags with oxygen in very high densities. It has always gone well previously because the trip takes just 30 minutes. But not this time around. A serious road accident had happened and the car with fish was trapped in a go-slow for three hours before it could move again. By that time, the fish were already dying. He lost 90% of them. The financial loss of approximately ₦ 200,000 was now gradually deducted from his salary!

Usually, customers come to pick up the fish. They are responsible for the containers to put the fish in. When the fish are small, less than 0.5 g, they can also be transported in plastic bags with extra oxygen. These bags are the responsibility of the supplier, which is the hatchery.

First of all, the fish should not be given any feed at least 12 hours before transport. After counting or weighing, the fish are put in a bag or container. But how many fish you can put in a bag or container? This question is not easy to answer, because it depends on the weight of the fish, the distance, the transport time, the weather conditions and the condition of the roads. See Photos 34 and 35 about transport and packaging materials for fish transportation.

Photo 34. A 50-l transport KEG for juveniles.

Photo 35. A transport 500-l insulated tank for juveniles or brood stock.

The transport time is actually more important than the distance. If the trip takes usually two hours, you have to prepare for safe transport for the fish for at least five hours!
If it is a hot day and you put the tanks or bags in a car without AC, you will risk the water temperature rising to more than 30 °C. As a result, the oxygen level of the water will go down and the fishes' rate of metabolism will increase. You create a big risk of getting very weak fish and mortality, either on arrival or when you put the fish in their new environment. If it is going to be a long trip, the best time to travel is in the evening and at night, if security allows you to do so. If the road is bumpy, the fish will get stressed and you will lose water on the way bit by bit.

Farmers usually stock between 500 to 1000 pieces of juveniles in the black containers showed on Photo 34. In the white plastic bags around 500 pieces of fingerlings are usually stocked. The bags are filled 1/3 with water and 2/3 with pure oxygen. So, not with air!
The supplier should always warn the farmer who buys the fish that the transport of the fish is the farmer's responsibility and dead fish on arrival cannot be compensated. If the supplier transports the fish, it is his responsibility.

The farmer who buys the fish, should have his pond well prepared, so that when the supplier arrives, the stocking of the juveniles or fingerlings will be fast and go smoothly. Don't start counting the fish again, they are already stressed enough!

After sales service

If you want farmers to be loyal to your farm, you must not only sell good quality fish but also offer after sales service. Inquire about the progress and growth of the fish and support the farmers with tips. Grow-out farms don't have the expertise hatcheries do, so you can certainly give them good tips and advice. The farmer's success will also be your success and if the farmer makes money with the sales of the table-sized fish, he will certainly come back to you for more fingerlings. Don't try to sell farmers quantities of fish they cannot accommodate because it just suits you to get rid of your juveniles. You will force your customer to sell his immature fish for a low price, because in this case the money for feed usually runs out before the end of the growth cycle.

8.8 HATCHERY SYSTEMS

I believe the combination of flow-through and recirculation systems is the best way forward for continuous supply of fingerlings and juveniles. The initial growth stages of at least the first three weeks, should be kept inside a building with constant water supply. Afterwards, the fry and fingerlings can be stocked outside in plastic tanks with periodic flushing, creating a green water environment.

8.9 GREEN WATER

I believe a green water environment is good for growing catfish in their first growth stages because of the following reasons:

- The algae in the water consume the waste of the fish and reduce the flushing rate without you having to deal with bad smelling water.
- The catfish feel secure because birds or other predators cannot spot them easily.
- It creates a favorable environment in which good bacteria have an advantage over the pathogenic ones.
- It creates a flora of organisms (natural food) which are beneficial to the growth of catfish fry.
- Green algae produce oxygen during the day but during the night they consume it and the oxygen level reduces. So, catfish need to have developed the labyrinth organ in order to breath oxygen from the air! I believe pathogenic organisms relying on oxygen, such as Aeromonas species, causing fin rot and elimination of the barbels, and Flexibacter columnaris, also called saddleback disease, are at a disadvantage in green water because they cannot flourish during the night.

- The catfish are usually healthy and don't suffer from bacterial diseases as much as in a clear water environment.
- Sick fish can even get over their illness without any treatment.

In order to keep a green water environment, you should never flush all the water at once, but in parts on a daily basis in such a way that the water is not getting too green, but also not getting to clear. The transparency of the water should be about 3 to 4 cm.

8.10 HYBRIDIZATION

In biology, a hybrid is the result of combining the qualities of two organisms of different breeds, varieties, species or genera through sexual reproduction. Since Clarias gariepinus (Photo 36) belongs to the genus Clarias and Heterobranchus longifilis (Photo 37) belongs to the genus Heterobranchus, this type of hybridization is called an intergeneric hybridization. In nature hybridization between two species doesn't occur, because they are reproductively isolated by morphological differences and different mating behavior. But when we proceed with an artificial reproduction of these species in the same way that we produce pure Clarias gariepinus, it results in viable hybrids called 'Hetero-clarias'.

The Hetero-clarias hybrid shows characteristics of both parental species, both in behavior and morphological appearance. Heterobranchus longifilis is the biggest species in the genus Heterobranchus. It has a dorsal fin with an adipose attachment on the back and very long whiskers. The meat has 30% more fat than Clarias has and is whiter in colour while Clarias meat is pinkish. You find both characteristics in Hetero-clarias.

The Hetero-clarias is not quite fertile and will not spend much energy on gonads. This affects the somatic growth and the FCR positively. However, its growth rate shows a lot of variation, so grading during the early stages of the growth cycle is very important to suppress cannibalism. The fish are hardy and do well in natural environments such as earthen ponds. Heterobranchus longifilis doesn't mature quickly compared to Clarias gariepinus. It takes close to three years in captivity. The most common cross breed consists of a Heterobranchus male with a Clarias female.

In the Netherlands one company specializes in producing the Hetero-clarias hybrid because of the white meat. They even branded the fish with the name 'Claresse'.

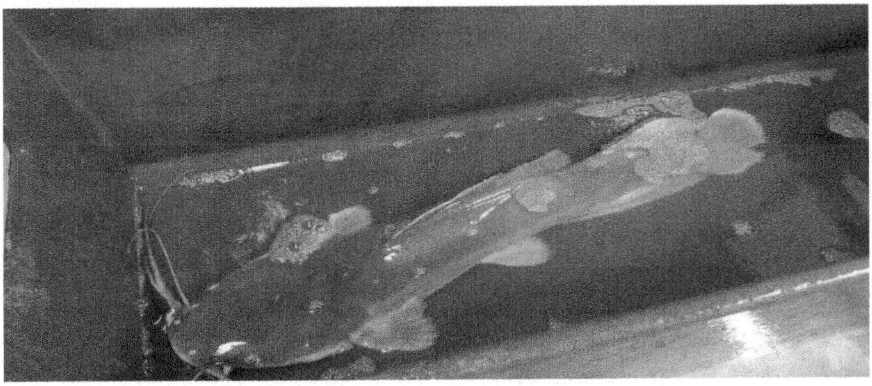

Photo 36. A Heterobranchus longifilis male.

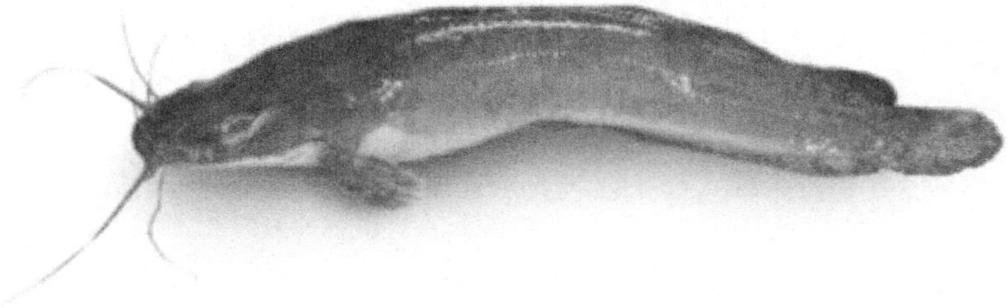

Photo 37. The Hybrid catfish 'Hetero-clarias or Claresse

8.11 DISCUSSION

Hatchery operations are not easy to manage and many farmers who start up hatcheries, face problems with the continuity of getting good results. So, at the end of the day growout farmers buying fish from these hatcheries, cannot rely on them, causing disruption in the production and profits completely. Therefore, many grow-out farmers start their own hatchery operations out of frustration.

Why is a hatchery so difficult to manage? Because running a successful hatchery depends on so many factors. When things go wrong, most of the time it is a multiple problem instead of a standalone one. What I have noticed in most cases is that the problem came down to lack of commitment of the hatchery manager and insufficient knowledge. They get results now and then, so in principle the facility works. But, if they get a good spawning and a good survival rate, they cannot repeat the results, because they don't know why things are going wrong or going well. Lacking good operational procedures, causes problems for repeating good results.

Hatcheries in Nigeria are not located in a controlled environment like in Europe, that's why you should adapt the procedures with the seasons. Running a hatchery is not only hard work, but also a way of life and an art form. You need to be naturally talented and have an eye for problems before they occur. As a young man I started hatchery operations in 1984 and till today I have not completely mastered the art of running a hatchery. That 'till today' has triggered me to stay involved in hatchery operations because it's a very challenging profession and never boring. Also, this work makes me a humble person and one who respects the law of God. I always say to my fellow farmers: "We don't make anything in a hatchery, we only support the law of nature created by God."

9

CATFISH DISEASES

1. INTRODUCTION
2. ENVIRONMENT
3. CATFISH AS A HOST
4. PATHOGENS
   1. BACTERIAL DISEASES
   2. PARASITIC DISEASES
   3. VIRAL DISEASES
5. MEDICATION
6. BIOSECURITY

9.1 INTRODUCTION

When I was studying at the University of Wageningen, I did an experiment with African catfish. I had to produce the fish for the experiment myself. This was my first step in becoming a fish farmer. The spawning was successful. I obtained about 40,000 fry to grow to the juvenile stage. At this stage the fish were stocked for the experiment. I needed only 2000 pieces. At the fingerling stage, I stocked the fish in two flow-through systems with crystal clear water and high oxygen levels. I believed this was the best water to grow my catfish, not knowing that this water actually stressed them and allowed virulent bacteria, such as Flexibacter columnaris, to develop very rapidly. The bacteria killed over 90% of my fish before my supervisor was aware of it. He quickly applied antibiotics which resulted in just enough fish to stock for my experiment. Actually, I learned more from growing these fish than from my experiment. My way of looking at fish has never been the same since! A song which relates to this experience is from The Beatles: 'Yesterday, all my troubles seem so far away!'

Fish diseases are the nightmare for every fish farmer. They can arise any time, even when you don't expect it. In most cases the losses are not severe when proper attention is paid, in other words diseases are manageable. But when not taken care of, a batch of fish or even worse a farm can be totally destroyed by diseases. They can even lead to bankruptcy of a fish farm. The life of fish is very dynamic, so many factors play a part in the outbreak of diseases. There are three major factors (Figure 23):

- Environment
- Catfish as a host
- Pathogens

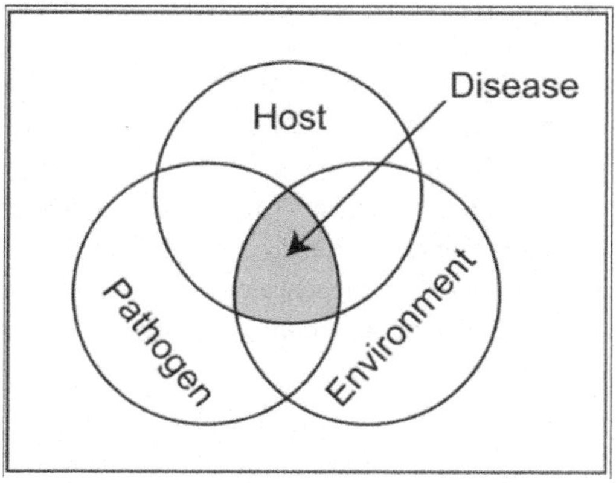

Figure 23. Illustration of interaction between host, pathogen and environment. (source: The nature of fish diseases by J.W. Warren)

As you can see in Figure 23, one factor can compensate another factor. For example, if the environment is not optimal for the fish, but there is no pathogen present, it will not be likely that the fish will get a disease. Or, the pathogens are present, but the condition of the fish is very good; the fish will resist the pathogen and will not get sick. The moment you abuse all three factors to a certain extent, the fish will certainly get sick. You also will understand from Figure 23 that when you treat the fish with medicine and they recover, it is actually a short-term remedy and not a sustainable solution. If you don't improve the three factors mentioned earlier, recurrence of the disease will be inevitable.

9.2 ENVIRONMENT

Fish live in water and at a fish farm they cannot escape from the water you provide for them, so it's your responsibility to give them water of a good quality. I believe that poor water quality is the number one reason fish develop diseases. I have seen cases where fish wanted to jump out of the water because of the poor water quality. Fish swimming towards or away from the inlet, are also signs that the water quality is bad.

Chemicals and factors causing a bad environment for the fish

The most common pollutants which cause a stressful environment for the fish are the waste products of the fish themselves and chemicals released by human activity.

Ammonia-nitrogen

Ammonia-nitrogen is a waste product produced by the fish and excreted through the gills. It is equivalent to the urine of mammals but more toxic; the levels of ammonia-nitrogen in the water should therefore be low.
Ammonia-nitrogen is actually a combination of ammonium and ammonia (NH_4^+ and NH_3: total ammonia-nitrogen or TAN), but only the ammonia is very toxic to the fish. The concentration of ammonia and ammonium in the water is present in a ratio, which depends on the pH of the water (Figure 24). When the pH is high, the percentage of ammonia is high and it starts to increase rapidly above a pH of 8. When the pH is low, the percentage of ammonium will be high, which is not a problem for the fish. But when the pH skyrockets, you have an emergency situation.

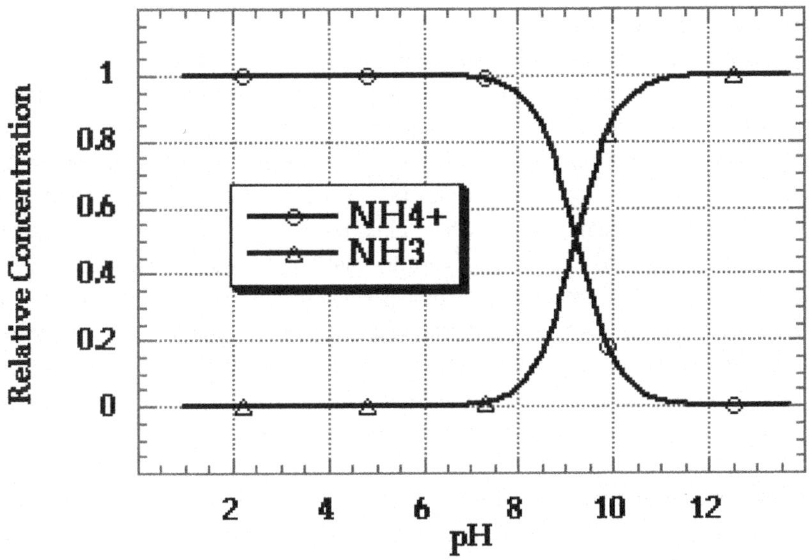

Figure 24. The link between ammonia and ammonium and the pH of the water. (source: Cornell Composting)

Levels of ammonia start to be toxic in the range of 0.1 mg/l for small catfish to 0.4 mg/l for bigger ones.

The signs of ammonia intoxication are:

- Loss of appetite; usually the first unspecific sign when a problem with the fish occurs
- Lethargy and changing colour
- Inflammation of gills
- Red barbels, red fins and inflamed eyes

If the incoming water is lower in ammonia concentration than the water in the tank, you will see the fish swimming towards the fresh water flow. The remedy is relatively simple: drain as much bad water out of the tank as quickly as possible and replace it with clean water without ammonia-nitrogen.
Test kits are available in the market to test the water for the ammonia-nitrogen concentration and the pH. You cannot see or smell the ammonia-nitrogen easily, so using your senses is not a reliable option!
For small fry and fingerlings an ammonia-nitrogen concentration below 5 mg/l, and for juveniles below 10 mg/l are recommended. For grow-out fish, the ammonia-nitrogen concentration should be below 25 mg/l at a pH of below 7.5

Suspended solids

Suspended solids are particles in the water which make the water cloudy. This turbidity can not only originate from waste materials, such as sand, laterite and clay, but also from the faeces produced by the fish or from an algal bloom. Catfish are not very sensitive to turbid water and thrive well in it. During the rainy season especially, the water can be very turbid because of soil erosion, but the fish will not face many problems. However, in the rainy season I have noticed greater mortality when using river water than during the dry season. I believe it isn't the solids suspended in the water that cause this problem, but the runoff from farms, containing pesticides and herbicides. Of course, in the hatchery you want to have clear water and when you use borehole water this will be usually the case. In general, high levels of suspended solids in the water will cause the oxygen level of the water to be low, because of a high biological oxygen demand (BOD). If the catfish have developed their extra breathing organ, they just compensate for this by taking oxygen from the air.

High levels of carbon dioxide

Carbon dioxide (CO_2) is not usually a big problem in African catfish farming, especially for all the growth stages of catfish bigger than fingerling size. In high density fish farming, the free carbon dioxide level increases in the water due to the metabolism of the fish. Normally, fish remove oxygen from the water and release carbon dioxide. African catfish however, release the carbon dioxide above the surface of the water so the carbon dioxide level of the water is not easily elevated. I want to draw your attention to this just for hatcheries, because they use borehole water which usually contains high levels of carbon dioxide (50 mg/l or more). This can easily be tested with a simple pH-test kit. Test the borehole water directly at the source without any contact with air. In most cases the pH will be slightly acidic, which means below 7. Take some of this water and fill a 1-l bottle with 100 ml (10%). Agitate the water by shaking the bottle for at least two minutes and measure the pH of the water again. Normally you will see an increase in pH to above 7. What happened is that the carbon dioxide, which is an acid, was released into the air and due to this process, the pH of the water rose (Figure 25).

Equation: $CO_2 + H_2O \longleftrightarrow HCO_3^- + H^+$ (H^+ is acid)

Carbon dioxide has a negative impact on the metabolism of young fish. They will behave sluggishly and refuse to eat. You can compare this to a church full of people with no ventilation; you start to feel sleepy. The concentration of free carbon dioxide should not be higher than 20 mg/l.

So, anyone who sets up a hatchery, has to degas the water from the borehole before it enters the fish tank. This can easily be done by a trickling tower above the water storage tank, which does not require extra energy, or by flowing the water into the tank through a showerhead.

	KH 2	KH 4	KH 6	KH 8	KH 10	KH 12	KH 14	KH 16	KH 18	KH 20
pH 7.8	1	2	3	4	5	6	7	8	9	10
pH 7.6	2	3	5	6	8	9	11	12	14	15
pH 7.4	2	5	7	10	12	14	17	19	21	24
pH 7.3	3	6	9	12	15	18	21	24	27	30
pH 7.2	4	8	11	15	19	23	27	30	34	38
pH 7.1	5	10	14	19	24	29	33	38	43	48
pH 7.0	6	12	18	24	30	36	42	48	54	60
pH 6.9	8	15	23	30	38	45	53	60	68	76
pH 6.8	10	19	29	38	48	57	67	76	86	95
pH 6.7	12	24	36	48	60	72	84	96	108	120
pH 6.6	15	30	45	60	75	90	105	121	136	151
pH 6.4	24	48	72	96	119	143	167	191	215	239
pH 6.2	38	76	114	151	189	227	265	303	341	379

Figure 25. The relationship between carbonate hardness or alkalinity, pH and free carbon dioxide (source: JBL Testlab).

Dissolved oxygen

In 1984 I started my farm in the Netherlands. Not only did I grow catfish, but I also tried a lot of other fish species. I was trading ornamental fish for outdoor ponds, including sturgeon (Acipenser gueldenstaedtii). I used to buy them from a farm in Eastern Europe and at the end of the summer, when the selling season was over, I just let the leftover sturgeon grow bigger for the next season at my indoor farm; by which time they would fetch a very good price! On Christmas day after feeding and checking the fish farm, I went to church and visited my family in the village. Coming back home in the afternoon, I experienced a disaster; the pump of the tanks where the sturgeon were stocked, had stopped and the water was not flowing through the tanks anymore. Most of my sturgeon had died because of oxygen depletion. I forgot to equip the system with backup aeration and the alarm system didn't work. Unfortunately, sometimes you have to learn it the hard way!

Low oxygen levels, also called environmental hypoxia, can cause mass mortality of fish. It is a very common cause of death at fish farms which farm non-air breathing fish, and also in rivers where factories dump chemicals into the water which consume all the oxygen. African catfish however, are hardy fish when it comes to oxygen concentrations in the water; they survive and even do well when the oxygen level is below 1 mg/l. Their organ to breath air is not immediately developed when they are born though. It takes about three to four weeks for the catfish to get to this growth stage. So, before this stage, oxygen levels have to be at least 4 mg/l. Therefore, hatcheries need running water in the tanks and/or aeration 24/7, and the tanks should not be overcrowded.

Oxygen levels in the water vary with the water temperature

Figure 26 shows that maximum oxygen levels reduce significantly when the water temperature rises. In addition, the metabolism of fish increases with increasing temperatures. As already stated, in a farm situation when a catfish has developed its air breathing organ, a low oxygen level is not a serious problem. However, it is good to realise that transporting fish in a car without air-conditioning, might lead to problems with maintaining an optimum water temperature, inducing mortality. So, for fish transportation ideally the water should not be more than 25 °C.

Herbicides and insecticides

Herbicides and insecticides are usually highly toxic for aquatic animals, including fish. Therefore, don't use insecticides at your fish farm to kill insects such as mosquitoes, flies or cockroaches. If they drop dead and fall into the water of your ponds or tanks, they can even be eaten by the fish! When spraying with herbicides on your premises to kill the grass, make sure that there is little wind so that none of the chemicals blow into the water of the tanks. Always be very cautious when using chemicals at your farm!

Another source of chemicals entering the water arises during the rainy season, when farmers use a lot of chemicals at their farm. Runoff flowing into the river might be contaminated with herbicides and insecticides. The dilution factor is high, but I believe even low concentrations of these chemicals can cause problems for your fish, if you use the river water directly. I believe the immune system of the fish is challenged and they will not be able to resist diseases as strongly as they can normally. The best way to use river water is to allow it to settle in a big reservoir with a retention time of at least two days.

During the dry season another threat occurs when fishermen use gammalin 20 (Lindane) to kill and catch the fish in the river. The flow of the river is slow and just a few litres of gammalin can kill a lot of fish. When you use this water to flush your ponds, you will certainly get problems with your fish. My advice is to always check the river and the water intake for floating dead fish; if so, delay pumping the water till you don't see dead fish anymore. The chemical will pass by if the river is still flowing.

Temp	SatDO		Temp	SatDO		Temp	SatDO
15	10.1		20	9.1		25	8.3
15.1	10.1		20.1	9.1		25.1	8.2
15.2	10		20.2	9.1		25.2	8.2
15.3	10		20.3	9		25.3	8.2
15.4	10		20.4	9		25.4	8.2
15.5	10		20.5	9		25.5	8.2
15.6	10		20.6	9		25.6	8.2
15.7	9.9		20.7	9		25.7	8.2
15.8	9.9		20.8	9		25.8	8.1
15.9	9.9		20.9	8.9		25.9	8.1
16	9.9		21	8.9		26	8.1
16.1	9.8		21.1	8.9		26.1	8.1
16.2	9.8		21.2	8.9		26.2	8.1
16.3	9.8		21.3	8.9		26.3	8.1
16.4	9.8		21.4	8.8		26.4	8.1
16.5	9.8		21.5	8.8		26.5	8
16.6	9.7		21.6	8.8		26.6	8
16.7	9.7		21.7	8.8		26.7	8
16.8	9.7		21.8	8.8		26.8	8
16.9	9.7		21.9	8.8		26.9	8
17	9.7		22	8.7		27	8
17.1	9.6		22.1	8.7		27.1	8
17.2	9.6		22.2	8.7		27.2	7.9
17.3	9.6		22.3	8.7		27.3	7.9
17.4	9.6		22.4	8.7		27.4	7.9
17.5	9.6		22.5	8.7		27.5	7.9
17.6	9.5		22.6	8.6		27.6	7.9
17.7	9.5		22.7	8.6		27.7	7.9
17.8	9.5		22.8	8.6		27.8	7.9
17.9	9.5		22.9	8.6		27.9	7.8
18	9.5		23	8.6		28	7.8
18.1	9.4		23.1	8.6		28.1	7.8
18.2	9.4		23.2	8.5		28.2	7.8
18.3	9.4		23.3	8.5		28.3	7.8
18.4	9.4		23.4	8.5		28.4	7.8
18.5	9.4		23.5	8.5		28.5	7.8
18.6	9.4		23.6	8.5		28.6	7.7
18.7	9.3		23.7	8.5		28.7	7.7
18.8	9.3		23.8	8.5		28.8	7.7
18.9	9.3		23.9	8.4		28.9	7.7
19	9.3		24	8.4		29	7.7
19.1	9.3		24.1	8.4		29.1	7.7
19.2	9.2		24.2	8.4		29.2	7.7
19.3	9.2		24.3	8.4		29.3	7.7
19.4	9.2		24.4	8.4		29.4	7.6
19.5	9.2		24.5	8.3		29.5	7.6
19.6	9.2		24.6	8.3		29.6	7.6
19.7	9.1		24.7	8.3		29.7	7.6
19.8	9.1		24.8	8.3		29.8	7.6
19.9	9.1		24.9	8.3		29.9 & 30	7.6

Figure 26. The relationship between the water temperature and the oxygen level at saturation (source: Fondriest Environmental Products).

Chlorine

Chlorine is a highly toxic chemical which is used for disinfecting materials and systems at the farm. It kills all the germs. But it is also used to disinfect the drinking water by chlorination. Fry are more susceptible to chlorine than adult fish. Lethal concentrations vary from 0.01 mg/l for fry to 0.25 mg/l for bigger catfish. Test kits for chlorine are rarely available in Nigeria, but they do exist.

What does chlorine do to the fish?

Chlorine is an oxidizer and kills every living tissue by causing necrosis. In catfish lethal levels of chlorine will cause the slime on the skin of the fish to disappear within 24 hours. The fish will certainly die. Also, chlorine attacks the respiratory system. So, the gills and the labyrinth organ will be affected too, causing breathing difficulties.

Fortunately, removing chlorine is not that difficult. Chlorine is relatively unstable in water and will vaporise from the water by itself within 24 hours, especially if you aerate it. A quick neutralisation of chlorine can be carried out with sodium thiosulfate at doses of 2 to 7 parts of sodium thiosulfate for 1 part of chlorine. If you have 2 m^3 water with a concentration of 0,5 mg/l chlorine, you need to neutralize the chlorine with 2 to 7 g of sodium thiosulfate.

pH and alkalinity of water

The pH of pure water is 7. In general, water with a pH lower than 7 is considered acidic, and water with a pH higher than 7 is considered basic. The normal range of the pH in surface water and ground water is between 6 and 8.5. This is also considered to be a good pH for fish to thrive in. The complete pH-scale ranges from 0 to 14. A pH lower than 6 and higher than 8.5 is dangerous for the well-being of fish. In both cases the slime on the skin and the gills themselves are affected. The slime on the skin actually dissolves. In addition, when the pH is high the ammonia - ammonium ratio is shifted towards the very toxic ammonia.

Alkalinity refers to the property of water to neutralize acid; it's an expression of the buffering capacity. For fish farming the alkalinity of the water should not be less than 25 mg/l equivalent calcium carbonate ($CaCO3$). Farming in extensive earthen ponds, creates an environment which enhances the natural production. When the water is acidic and the alkalinity is low (less than 25 mg/l), the buffer capacity of the water should be enhanced. In Nigeria this is usually carried out by adding lime to the water or onto the bottom of the pond. The dosage is 300 kg per 1000 m^2 of the pond.

This way algae can grow and the natural cycle of phytoplankton to zooplankton is triggered. Catfish thrive better in a pond with an algal bloom than in a clear water environment. When the pH of the soil and the water is already above 7.5 at the end of the day, there is no need for additional liming! New ponds which just have been constructed, need lime to make them fertile: 2000 kg for sandy soil and 4000 kg for clay soil per ha. This can be carried out in an empty pond before filling it with water. The lime also disinfects the soil and enhances mineralisation of the mud. One or two months after the ponds have been filled, you can test the total alkalinity with a test kit available in Nigeria. It should be more than 25 mg/l. If the alkalinity is less than 25 mg/l, you can add another 200 kg of limestone/ha, and repeat this as long as necessary (source: FAO, pond conditioning through liming).

Heavy metals such as copper, zinc, cadmium, lead

Some heavy metals, in small quantities though, are a necessity for growing fish. They are part of the essential micronutrients which are even added to the feed. But when they occur in the water, even at low concentrations, they will be toxic to the fish and should be avoided by all means.

Normally in groundwater you will not have problems with heavy metals, but when you are using copper or galvanised pipes, you certainly will elevate the levels of copper, zinc and cadmium in the water. This will be devastating for the results of your hatchery. If you use river water, you will not usually have problems with these heavy metals, but when a river passes heavy industrial or urban areas, you have to be cautious. Fish can become chronically sick and get susceptible to diseases or die directly from this type of pollution. Most of these pollutants damage the internal organs and eventually this will lead to mortality. Fish eaten from polluted rivers are even dangerous for human health. The natural food for river fish is polluted as well, and the fish will deposit these heavy metals in their organs and fat.

Concrete

Newly built concrete ponds filled with water, will elevate the pH (basic) of the water for some weeks, but after this period the pH will be adjusted to the pH of the incoming water. So, it is advisable to leave a pond at least one or two weeks filled with water without fish in order to neutralize the effect of concrete. After two weeks it's usually safe to stock fish in it, but flush the water first and refill.

Water temperature

Catfish are tropical fish which cannot survive in cold water. The water temperature should be between 25 - 30 °C for good growth and health. The optimum temperature is 28 °C. A short period of one or two hours of 12 °C will not cause the catfish to die. If the water temperature increases gradually after the short cold period, they will just wake up again with little problem. However, longer periods of cold water might initiate fungal growth on the skin, resulting in a quick death. Temperatures above 34 °C are also dangerous for catfish, because this temperature comes close to the level at which proteins in their body start to denature. Crucial enzymes and metabolic processes especially are irreversibly damaged leading to mortality as well.

Monitoring the water quality

In order to find out if the levels of ammonium-nitrogen, nitrite, pH, iron, hardness, temperature etcetera are within limits for growing fish, you need to have several test kits or all of them combined in one box (Photo 38).

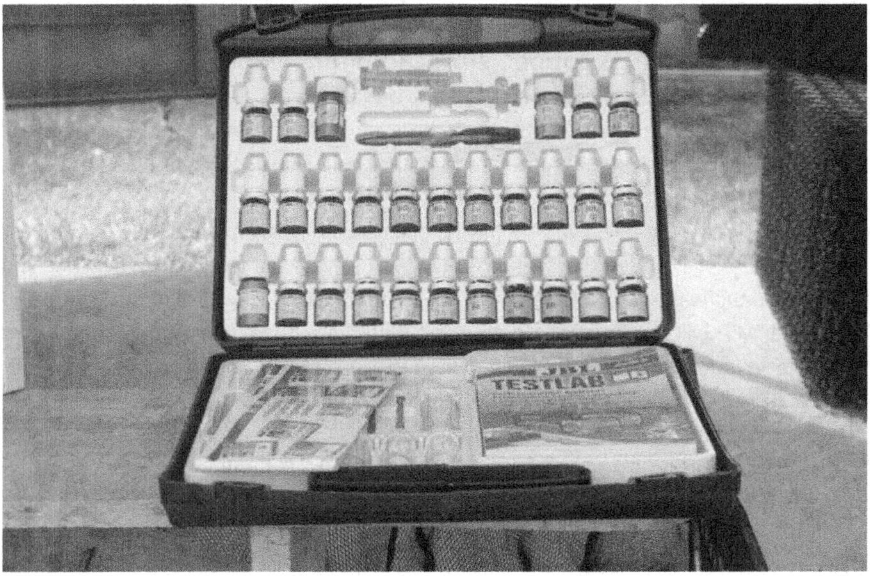

Photo 38. A box with several test kits for monitoring the water quality.

These test kits are very simple to use and have a very good instruction manual. They have been especially developed for fish farmers and owners of ornamental fishes.

Feed-related deficiency

If the fish at your farm are completely dependent on the feed you are providing, the feed should provide all the requirements for the fish. So, the levels of protein, fat, ash, minerals and vitamins should be balanced in order to allow the fish to obtain optimal health and optimal growth. Depending on the growth stage, the requirements change and should be adjusted. If the feed contains too little of a certain requirement, this will immediately be shown by retarded growth and limited feed intake. It becomes even worse if some requirements are completely missing, such as vitamin C. This will cause the fish to develop a vitamin deficiency called 'broken head disease' and they will eventually die from it, if not taken care of.

If you don't make your own feed, there is not much to worry about, as long as you are buying the feed from a respected feed mill with a proven track record. As already mentioned earlier, I don't believe in making your own feed. That's why I will not go into this subject in detail.

9.3 CATFISH AS A HOST

The quality and the condition of the fish you stock in your ponds or tanks, are two underestimated factors in prevention of diseases. Strong healthy fish can put up with a lot! When you buy fish from hatcheries to stock them in your grow-out ponds, you will find out that fish from different hatcheries will yield different results. First of all, the growth is different, which means that also the health of the fish varies too. Are the fish well fed and given the best feed at the hatchery? Has the brood stock been selected well and also given the best feed? The genetics of the fish also play a part. Well-domesticated lines of catfish will adapt better in a captive environment and will not feel so stressed as wild undomesticated catfish. The fry stage is crucial when it comes to development of the intestinal system and bone formation. When developmental problems arise, they will not fade away, but remain till the fish are harvested at table size. It's like building a house on a good or a bad foundation. Some hatcheries are using antibiotics preventively during the entire growth cycle. This will definitely weaken the immune system of the fish.

Probiotics in fish feed

In the future diseases will undoubtedly become more problematic. Strains of pathogenic bacteria are becoming resistant to the most commonly used drugs. So, increasing the health status of the fish is a good approach in order to avoid diseases and it is also environmentally friendly. Probiotics applied to feed to enhance the general health is already common practice for land-based animals, but are recently also introduced in aquaculture. Probiotics are harmless bacteria which contribute to the intestinal microbial balance. They operate in several ways:

- Competition between these 'friendly bacteria' and pathogenic bacteria, stabilises the intestinal tract.
- Probiotics produce compounds which inhibit the growth of pathogenic bacteria.
- Probiotics enhance the immune system of the fish.

The use of and ways to incorporate probiotics in fish feed are relatively new and more research is needed to be sure they are beneficial to the species of fish they are given to. This development is certainly promising and worth mentioning.

9.4 PATHOGENS

Like humans and other animals, fish suffer from diseases too. The fish's defences against diseases are specific and non-specific. Non-specific defences include skin and scales, as well as the slime layer secreted by the skin, which traps microorganism and inhibits their growth. If pathogens breach these defences, fish develop inflammatory responses that increase the flow of blood to the infected areas, resulting in the delivery of white blood cells which attack the pathogens.

Stressed or sick fish can often be recognized by abnormal behaviour, such as decreased appetite, nervous or unsteady swimming, remaining in a vertical position at the surface of the water, or by clinical symptoms, such as mutilated barbels or fins, dropsy, white or red-brown spots on the skin, pop-eyes etcetera. However, keep in mind that these symptoms are not specific, so, further laboratory tests may have to be conducted.

At the farm you should be monitoring daily for diseases, especially during feeding when catfish usually swim up to the surface of the water. For proper diagnoses of bacterial, fungal and parasitical diseases, swab preparations of the skin, the gill filaments, intestines etcetera should be made and examined under a microscope with appropriate magnification. When a swab does not show any parasites, you most likely have bacterial problems to deal with. In this case, you should try to culture them on an agar plate and run an antibiotic sensitivity test. After diagnosis, specific therapy may be administered. Since most farmers have no laboratory facilities to culture bacteria, they just decide to apply a broad-spectrum antibiotic. It takes time to get results after setting up a culture and in the meantime your fish are dying. So, you are also under time pressure to use the right antibiotic to get a quick improvement in the health of your fish.

Diseases can be caused by 3 types of pathogens:

- Bacteria
- Parasites
- Viruses

This handbook does not cover all the possible diseases in fish, but focuses on those diseases which usually occur in African catfish.

9.4.1 BACTERIAL DISEASES

The majority of bacteria which cause diseases in catfish, are normally present on the surface or in the gut of the fish, or in the water of their culture environment. They cause diseases when the catfish are stressed.
When you carry out a bacterial examination on a fish with a bacterial disease, in many cases you will find out that the disease is multi-bacterial and that the disease is caused by many different bacterial species. The signs of different bacterial diseases in fish are also non-specific. So, to be sure about which bacterial species has caused the problem, is almost impossible either after a physical examination or after culturing the bacteria in a laboratory.

Flexibacter columnaris

This is a Gram-negative aerobic bacterium which causes infections on the mouth, barbels and back of the catfish. Therefore, it is also known as 'saddleback disease' or 'cottonmouth disease'. The bacteria also produce toxins causing the fish to swim erratically. Strangely enough these bacteria thrive well in well oxygenated water with a very low organic load. You will always find the bacteria in the water, but when the situation is stressful for the catfish and/or when new catfish are introduced into clean water, the bacteria might get a foothold and cause a disease outbreak. Hard water also causes the bacteria to be more pathogenic.

In a green water system, as I discussed in chapter 8, it is not likely that catfish will get this disease.

The disease can be treated with industrial salt at a dosage of 6 kg/m^3 as a bath treatment. This can also be combined with oxytetracycline (OTC) at a dosage of 30 g/m^3. The OTC and salt treatment should be repeated after the second day, if the concentration of salt has significantly been reduced by flushing.

It is important to start the treatments as soon as you notice the first signs of the disease. If you don't, you will be dealing with a virulent strain of Flexibacter columnaris, which can kill almost all your population in the tank within two days. Smaller catfish are more susceptible than bigger ones.

Aeromonas hydrophila

Aeromonas hydrophila is a heterotrophic Gram-negative bacterial species, mainly found in warm climates. The disease is known as 'ulcer disease' or 'red sore disease'. It causes lesions on the body, and can also cause septicaemia by invading the organs of the fish.
The bacteria primarily affect freshwater fish such as catfish, tilapia and many other tropical fish. Other symptoms are lack of appetite, swimming abnormally, pale gills and a bloated

appearance. To be sure you are dealing with Aeromonas hydrophila, run a laboratory test by taking swabs from organs and lesions and grow them on a blood agar. A sensitivity test for various available antibiotics to enable application of the most effective ones, can also be included. Aeromonas hydrophila can thrive in low and high oxygen water and prefers organic polluted water.

Catfish are not usually affected by this bacterial species, but at an increased stress level caused by poor handling, bad water quality, poor feeding, temperature fluctuations etcetera, it might break out. In Nigerian catfish farming this disease is the most common one to affect farms. Usually it doesn't lead to mass mortality, but it can reduce a population in a tank by 10 to 20%. It starts with a few fish dying in one day and increases to 1 - 2% of dead fish within the next 24 hours. Gradually the mortality reduces again till it totally stops. When the fish are in a good environment, the disease is usually over after 10 days.
When you have a disease outbreak in bigger fish, you will see a lot of slime and skin particles making the water cloudy. You have to flush more in order to keep the water clean.
Treatment can be applied in the same way you treat a disease caused by Flexibacter columnaris: 6 kg/m^3 industrial salt and 30 g/m^3 OTC. Apart from this, you could also apply OTC in the feed at a rate of 50 g/ton of fish a day for a period of 10 days.

Potential concern for human health

An Aeromonas hydrophila infection can be spread from fish to humans and vice versa. If you are healthy, it will not be likely to happen, but small children and people with an immune deficient system can get infected. The symptoms, such as diarrhoea, are the same as in cholera. I always recommend fish farmers to feed the fish with a spoon or a basket and not by hand, because this disease can also be spread from human beings to fish. Sick fish should be handled with gloves!

Citrobacter freundii

As with the other bacterial diseases, Citrobacter freundii, which is a Gram-negative motile bacterial species, is triggered by bad environmental factors, such as poor water quality, high water temperature and other stress factors. A typical symptom for infected African catfish, is a swollen head kidney. This can be seen just after the pectoral fins. It can even burst through the skin leaving an ulcer behind. The fish are sluggish and will certainly die from it. The head kidney is a blood forming organ and if it is affected, it will affect the whole body causing septicaemia. This occurs when a bacterial infection elsewhere in the body, for example in the kidneys, enters the bloodstream. This is fatal in most cases because the bacteria and their toxins can be carried through the bloodstream to all the other organs of the fish.
Once when I had a problem at the farm with this disease at the growth stage of juveniles, we ran a sensitivity test which indicated that Trimethoprim as a bath treatment at a dosage of 10

g/m^3 performed well. It can also be used in the feed at a dosage of 30 g/ton of fish in a period of 7 days. Trimethoprim is also considered a broad-spectrum antibiotic

Gram-negative bacteria: what does it mean?

In microbiology, laboratory tests are run to find out which type of bacteria is causing the fish to get sick. Gram staining is a method for bacterial differentiation: if the cell retains crystal violet, it is called Gram positive-purple coloration, if it doesn't, it is called Gram negative-red or pink coloration). Hans Christian Gram invented the Gram staining method as early as 1882.

The effectiveness of antibiotics in treating bacterial diseases

Unfortunately, the use of antibiotics is more an exercise of trial and error. While your fish are dying, you don't have the time to go to a laboratory to find out which bacterial species is responsible for the disease and which antibiotic will be effective. Also, in Nigeria you rarely find laboratories which specialise in fish diseases and are able to run a sensitivity test. A sensitivity test goes as follows:

1. Isolate the bacteria from a wound, ulcer, skin or the internal organs of the infected fish.
2. Spread and grow the isolated bacteria on an agar plate (a bacterial growth medium solidified with agar in a Petri dish).
3. Take the dominant growing bacteria colony to another agar plate which contains several types of antibiotics in little circles. The bacteria will grow within the antibiotic circles, but when the outside of one of the circles shows no bacterial growth, it means that the bacteria are inhibited or killed by that type of antibiotic.

When you have found the right antibiotic by a sensitivity test, the veterinarian can prescribe the one you should use to treat your fish effectively.

The silver bullet

As already mentioned, treating bacterial fish diseases with antibiotics is in many cases a matter of trial and error. However, there is this broad-spectrum antibiotic Enrofloxacin, which is considered to be very effective for bacterial diseases in fish. The brand name in Nigeria is Kenflox 20% Oral. When nothing works, this medicine will be your last hope. That's why it is also called 'the silver bullet': a magical solution to a confusing problem! This antibiotic can be applied in the water or in the feed. In the feed the recommended dosage used is 10 mg Enrofloxacin/kg fish a day for 5 days. Added to the water, farmers use a dosage of 5 ml for 1 m^3 of water which effectively is not more than 1 ml of 100% Enrofloxacin per m^3 of water, and they repeat the dosage for 3 days.

9.4.2 PARASITES

Introduction

Parasites are easily spotted on fish either by eye or microscope, depending on the species. In order to multiply, some parasites need a secondary host, such as a snail or a bird. These kinds of parasites you will rarely find at indoor farms, but only at outdoor farms with ponds in a natural environment, filled and flushed with river water. Other parasites don't need a second host. They can be found indoors as well as outdoors. Parasites in catfish are not common and they have little effect on catfish bigger than 50 g. However, in hatcheries parasites can cause severe damage and if not treated in time, you will lose the majority of the young fish within a week.

The parasites spotted on catfish are ecto-parasites, and are based on the skin or gills. They are easy to scrape off with a swab or surgical blade and put on a glass slide for identification under a microscope. The list of parasites is long. Therefore, I will only mention the parasites I have actually seen and experienced in catfish myself.

Dactylogyrus – Gill fluke or gill worm

In 1994 I wanted to refresh my brood stock at my farm in the Netherlands, to avoid inbreeding. I bought matured catfish from a farm which had formerly been used for eel production; a massive farm with an enormous filtration system. The brood stock I bought was infected with gill worms, also called gill fluke or Dactylogyrus. The transmission of the gill worms to the catfish was caused by the eels which had previously been grown. Although there were no eels at that farm anymore, eggs must have stayed behind in the filtration system and they were able to find the catfish. Big catfish don't mind gill worms and I didn't put my new brood stock in quarantine and neither didn't check if they had any ectoparasites. So, when I started to use these brood stocks, I transferred the gill worms to my freshly hatched fries. The signs were clear because the young fish behaved sluggishly by hanging in the upper layers of the water column. This was confirmed when I used my microscope and saw them hooked to the gills of my little fry. Next, I used a medicine called Masoten (organophosphate insecticide) to kill those worms. This medicine is widely used in eel farming, but also on animals, such as cows and goats and even human beings. The veterinarian guaranteed it was a very safe and effective medicine to use. So, I treated all my fry stocked in more than 10 tanks at once. After two days, all the fry were dying and I discovered that catfish cannot handle this medicine; it's toxic to them. With the next batches of fry, I disinfected the systems thoroughly before hatching, and if they got sick from gill worms, I just allowed it to happen because they didn't die from it and finally grew over it. As a young fish farmer, I was crying over my fish and over my mistakes but it didn't kill me, and what doesn't kill you makes you stronger!

Multiple mistakes

- I didn't test the brood stock for parasites!
- I didn't put the brood stock in quarantine!
- I didn't test the drug on a small quantity of fish!

Figure 27. Drawing of the fish parasite Dactylogyrus.

The Dactylogyrus (Figure 27) is a parasite which is called a monogenetic trematode and varies in size between 0.05 and 3.0 mm. They have large grappling hooks which are used to attach themselves to the gills of the catfish. This irritates the gills usually causing the fish to jump and rub themselves against the walls of the tank. Infected catfish of two to four weeks old hang at the upper layer of the water column and show reduced eating behaviour, but fortunately they don't die of it. If the catfish fry are taken good care of, they actually grow over it. When they reach the juvenile stage, the parasites on the gills will have vanished almost completely.

Note. The eggs of the Dactylogyrus are not affected by the medicine used to get rid of the gill worms.

Gyrodactylus – Skin fluke

The fish parasite Gyrodactylus (Figure 28) belongs to the same family as the Dactylogyrus. Gyrodactylus parasites attach themselves to the skin of the catfish and bear live young. In an adult Gyrodactylus you can already discover the babies in the body, if you put a swab of it under the microscope. So, they multiply very quickly and can turn into a big problem for catfish smaller than 20 g, if the conditions for the Gyrodactylus are favourable. In addition, the consumption of the skin of the catfish by this parasite favours the introduction of bacterial infections on top of it. The catfish can die in large numbers when not taken care of.

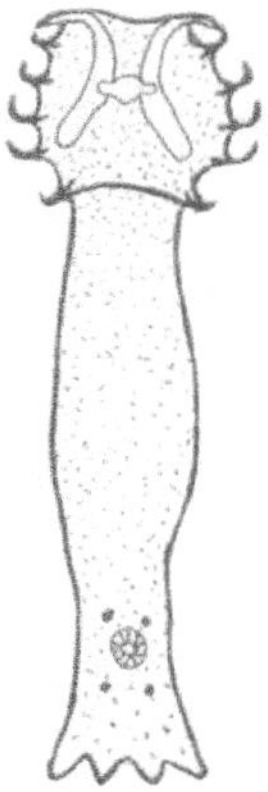

Figure 28. Drawing of the fish parasite Gyrodactylus.

Fortunately, formalin is an effective medicine to stop the invasion. The dosage is 28 ml/m^3 at the concentration of 38% formalin. Use the formalin in combination with salt at a dosage of 1 kg/m^3. The water should not be flushed for 12 hours and afterwards start flushing gradually. Checking with the microscope, you will notice a close to 100% reduction of the parasites. If needed, repeat the treatment the second day.

Note. Formalin is also very toxic for catfish and the dosage used for treatment is not far away from their lethal doses as well! Calculation of the exact volume of water should therefore be taken seriously. If you are not sure whether the fish can take the dosage, test it on a small number of fish!

For example:
The fish pond has the dimension of 5.5 x 2.4 x 0.8 m (length x width x depth). The volume is 5.5 x 2.4 x 0.8 m = 10.56 m^3. The dosage is 10.56 x 28 ml/m^3 = 295.68 ml

- Buy the formalin from a trustworthy supplier so that the % of formalin is according to the label on the bottle and not diluted. 38% formalin has a very penetrating smell, so use it with care. Wear gloves and a face mask.

Protozoan parasites

Protozoan single-celled parasites are troublesome and can multiply even more quickly than the parasites I have mentioned above. They multiply by binary fission: one parasite becomes two, two parasites become 4 and so on.
The ones I have seen on African catfish are:

Chilodonella, Trichodena and the white spot disease-causing Ichtyophthirius multifilis

All of the single-celled protozoa, such as Chilodonella, Trichodena and Ichtyophthirius multifilis (Figure 29) are categorised as ciliates and cause grey thick slime on the skin or white spot disease. Ciliates belong to a group of protozoans which have hair like organelles around the cell, called 'cilia'. By taking some slime from the skin, all of these parasites are easily identified under a microscope at a magnification of just 100 times. You can even see them moving. They stick to the skin, fins or gills of a catfish, causing either grey thick slime on the skin, or, in case of white spot disease, hundreds of spots of circa 0,5 mm to 1 mm on the entire body of the fish. Always use live fish for examination, since many parasites leave the host very quickly after the fish dies. By feeding on the gills, skin and slime of the fish, these parasites cause a lot of damage.
White spot disease has a life cycle: the larval stage is found on the fish, but after feeding and growing to a spot of approximately 1 mm, the parasites will leave the fish and mature. They multiply at the bottom of the tank and produce hundreds of new small larvae (cysts or spores) which in their turn look for the weak fish to infect. The life cycle takes six to eight days and is highly dependent on the water temperature.

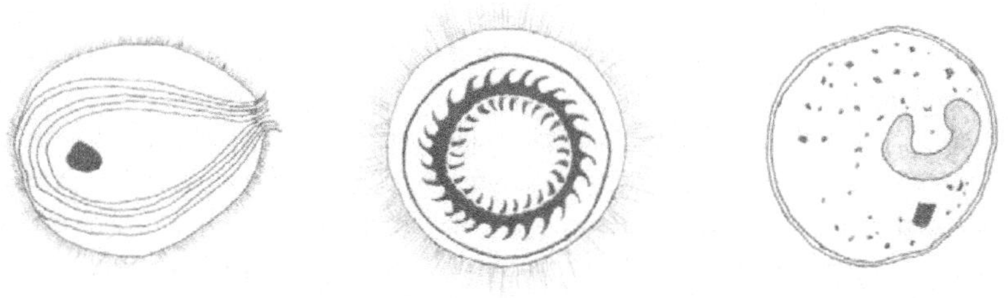

Figure 29. Left to right: single-celled catfish parasites Chilodonella, Trichodina and Ichtyophthirius multifilis.

Treatment against protozoan parasites

One treatment usually doesn't solve the problem with protozoan parasites. White spot disease especially needs multiple treatments to eventually stop the disease completely. Only the free-swimming spores of Ichtyophthirius multifilis are sensitive to chemicals, the other stages are not. Dead fish should be removed swiftly in order to reduce the spread of the disease.

Many chemicals can be used to kill the parasites:

- Salt (NaCl)
- Formalin
- Acriflavine
- Chloramine T
- Methylene blue

Since not all the medicines are easily available in Nigeria, I 'll focus on the treatments of protozoan parasites using salt and formalin.
Chilodonella and Trichodena can be successfully treated with 28 ml/m^3 formalin (38%) in combination with 1 kg salt/m^3 for 12 hours, followed by 2 to 3 treatments every other day. White spot disease can only be treated with salt as a permanent bath of 7 kg/m^3 for at least 5 days. It is important to start with the treatment at an initial stage of the disease! When catfish fingerlings and juveniles are infected with these parasitic diseases and not treated in time, the result can be 100% mortality.
The population of parasites you will find after treatment, should be close to zero, when checked with a microscope. The catfish should eat better and behave normally.

9.4.3 VIRAL DISEASES

Definition of a virus

A virus is a small particle that is capable of infecting a cell and has the potential to cause a disease. The cell that is infected by the virus is called a 'host cell'. Most viruses are too small to be seen with a normal microscope. A virus is unable to reproduce without the help of the host cell.
There are a few viral fish diseases, but till today researchers have not yet isolated any virus in African catfish which causes serious damage. In general, viral fish diseases are not treatable and can cause a high mortality especially among young fish. Fish which survive the disease, become carriers and infect other fish which have not yet been exposed.

New fish viruses are still being diagnosed, which is a serious threat to aquaculture in general. Movements of fish from one continent to the other, make it even easier for viruses to spread over the world very quickly. One of the latest examples is the Tilapia lake virus (TilV) which was reported in 2014 in Ecuador and Israel. At present it has been reported in Asia and also in Lake Victoria among wild populations. It was suspected of causing the mass mortality at some of the cage farms in Volta Lake, but that did not appear to be the case.
Ghana authorities have strict regulations when it comes to importation of improved tilapia strains into the country; actually, farmers can only use the local indigenous strains.

However, this holds back the introduction to Ghana of good growing strains of tilapia, produced in other parts of the world. But at the same time these regulations at least help the biosecurity and avoid the introduction of diseases which can cause severe damage to both wild and farmed tilapia populations.

9.5 MEDICATION

A good farmer should have the most important medicines in stock, well stored in a cool, dark and dry environment. Fish diseases spread very quickly and once detected, the fish should be treated immediately. It will save you a lot of fish and money if you don't waste time!

There are 3 ways to apply medication:

- **Apply in the feed**

Applying drugs in the feed can only work if the fish are still eating. Usually the dosage is given as x mg/kg of fish a day. The feed is given in % of bodyweight and should be reduced to less than 1% of the bodyweight in a day. So, you need to know the number and the approximate size of the fish in the pond you want to treat with medication.

Example 14. How to calculate the daily dosage of medication applied in feed?
Suppose that you have stocked 5000 catfish and after 2 weeks mortality starts and you suspect a bacterial infection. The fish are stocked at an average of 10 g a piece and in 2 weeks they were fed 40 kg.
The estimated biomass of the fish will be:

5000 x 0.01 kg + 40 kg x 1.0 (FCR=1) = 90 kg

The dosage of applying Oxytetracycline (OTC) in the feed is 75 mg/kg of fish.
So, the dosage on a daily basis for a period of 7 days is:

90 kg x 75 mg/kg = 6750 mg or approximately 7 g.

You feed the fish 1% of the bodyweight on a daily basis which is 90 kg x 1/100 = 0.9 kg

In summary you apply 7 g of OTC per day in 0.9 kg of starter feed for 7 days.

Note. Mixing the antibiotic in the feed should be carried out with gloves. Normally the antibiotic will stick to the oil-coated starter feed. If the feed is too dry, you can add a little vegetable oil to make the feed sticky for the antibiotic.

- **Apply to the water**

Applying medication to the water is not always feasible. When fish are stocked in a pond with a very large volume of water, it will simply cost too much. In some cases, you can reduce (lower) the volume of water in order to save on the amount of medication. You can give the fish a permanent bath by applying a dose of medicine in the water and let the fish remain in it for hours, or you can give the fish a dip treatment by applying a dose which is so high that you only dip the fish in it for a minute and take the fish out again. In the case of a treatment with antibiotics, a permanent bath is usually given. Formalin and salt treatments can be carried out as a permanent bath at low concentrations or a dip treatment when high concentrations are applied. I personally don't like the dip treatment except for cleaning the brood stock before using them for spawning or before introducing them to other fish.

Example 15. How to calculate the daily dosage of medication applied in water?

A concrete tank has the dimension of 5 x 2 x 0.8 m and you want to treat the fish with OTC 40 g/m^3. You reduce the water level to 0.4 m. How much OTC should you use?

Solution:

5 x 2 x 0.4 m = 4 m^3 volume of the pond. Dosage: 4 m^3 x 40 g/m^3 = 160 g.

The OTC should remain in tank for at least 4 hours, followed by adding fresh water.

Note. OTC is yellow, but when it starts to decompose after some time it becomes brown. So, the water colours brown too, but that is not a problem. There is no need to flush it.

OTC can be bought in different grades of purity; all the dosages given are for the active ingredient in the medicine.

Example 16. How to calculate the dosage of medication based on the percentage of active ingredient?

The OTC you are buying has an active ingredient of 100 g OTC/kg, which means that the medicine has 10% active ingredient.

The dosage of 40 g/m^3 has to be multiplied by 10 = 400 g/m^3.

Always try to buy medicine with 100% active ingredient.

The inactive ingredient will pollute the water.

- **Apply by injection**

Applying medicine by injection is not used in catfish farming, but it is normal with ornamental fish which are very expensive or which have emotional value. For catfish It would be just too much hassle and expense to inject each fish separately!

9.6 BIOSECURITY

Biosecurity in fish farming consists of practices that minimize the risk of introducing an infectious disease and spreading it to other fish, either at the farm or outside the farm. Prevention of diseases is always better than curing them. My experience is that a disease which has the chance to develop at a farm, is not easy to eliminate.

Measures to take in order to have a bio-secured area:

- Buy your fingerlings, juveniles or brood stock from a known farm, visit it and check the supplier is taking biosecurity seriously.
- Keep the fish which you buy from an unknown source, in quarantine for at least a couple of weeks. These fish should not be in contact with the stocks of fish which are already at your farm.
- Customers who visit your farm to buy fish, shouldn't be close to the fish pond but should be directed to a point of sales. Customers go to many farms and their equipment and baskets are very dirty most of the time.
- After using your equipment, such as graders, baskets and catchers, always clean them with a high-pressure cleaner and disinfect with a 1% chlorine solution afterwards.
- Before using equipment, always wash and rinse it very well.
- Never feed the fish with your hands.
- Remove dead fish in the morning and evening with a catcher, but avoid spreading disease from one pond to another.
- Use a separate catcher for tanks with diseased fish and ensure they are properly disinfected before using them in another pond.
- Eliminate rodents.
- Keep the environment clean; waste disposal has to be carried out on a weekly basis.
- Preferably don't let people who work in the hatchery, handle table-sized fish.
- Eliminate dead fish outside the compound to prevent birds, rats and other animals eating them and contaminating your farm.
- Don't sell dead fish which have been taken from a pond.
- Staff should understand the principles of biosecurity and should be lectured about it on a timely basis.

10

FISH FARM MANAGEMENT

1. INTRODUCTION
2. HOW TO MAKE A PRODUCTION PLAN: CASE STUDY
3. CHECK POINTS FOR THE MANAGER

10.1 INTRODUCTION

Fish farm management starts with making a production plan and following it so that the production estimated will be achieved. For farm management records are very important; they show the progress of the farm and whether you are on track to reach your targets.

10.2 HOW TO MAKE A PRODUCTION PLAN: CASE STUDY

Let us start with an example of a catfish farm with a flow-through system. I'll take you through the necessary calculations to estimate the production and the inputs needed.

Quantify the facility:

- 50 concrete tanks of 50 m^3 each
- 15 tanks for pre on-growing, 11 m^3 each
- Hatchery with a capacity of 1,000,000 juveniles
- Borehole of 5 m^3/hour
- Intake from the river with a capacity of 200 m^3/hour
- 100 kVA generator and 11 kVA generator

Growth cycle and production figures:

- Hatchery producing juveniles: 2 months
- Pre on-growing producing post-juveniles: 45 days
- On-growing producing table-sized fish: 150 days
- Density in pre on-growing: 500 pieces/m^3
- Density in on-growing: 100 pieces/m^3
- Water usage: 1.4 m^3/kg of feed
- FCR (feed conversion rate) for pre on-growing: 0.7
- FCR for on-growing: 1.25

Target: table-sized catfish in the range of 1.5 to 2.0 kg

Calculation

We will start with calculating the maximum production for the on-growing stage. Then we will calculate if the facilities for the pre on-growing stage and the hatchery stage are sufficient to provide the fish needed for the on-growing stage.

Each tank in the on-growing system can provide a minimum of 2 cycles/year: 365/150 = 2.4 cycles, but we reduce it to 2, since the tanks will be empty for some time before restocking and in case the sales of the table-sized fish are delayed a bit. It is always good to allow yourself some space to manoeuvre. Be realistic.

Production capacity in the on-growing system

50 m^3 (volume of pond) x 100 fish/m^3 (density) x 1.75 kg (average weight) x 2 (number of cycles/annum) x 50 (number of ponds) = 875,000 kg of table-sized catfish per annum. In practice, you will sell 2 ponds every week with only a 2 week's break! The production of each pond is 8,750 kg.

Post juveniles needed from pre on-growing per annum:

Post-juveniles needed per annum: 100 x 5000 x 1.1 (+/- 10% mortality during 150 days) = 550,000.

Water needed per annum:

1.4 m^3/kg feed x 1.25 (FCR) x 875,000 kg (total production) = 1,531,250 m^3 or 4,200 m^3/day.

Advice

The intake pump provides 200 m^3/hour. In order to supply 4,200 m^3/day, it has to pump for 21 hours. However, if the pump fails, you actually do not have any back-up in the reservoir. If you want to have a back-up water supply for 1 day, the reservoir should have a volume of at least 4,200 m^3 and the pump should have a capacity of 400 m^3/hour. After one full day of pumping, you will have served the need for one day and have obtained circa 4,200 m^3 extra for the following day. If your reservoir is already full, you only need to pump approximately 11 hours a day to keep it that way.

Production capacity in the pre on-growing system

We have 15 tanks for the pre on-growing which takes 45 days. We can do 365/45 cycles a year = 8 cycles a year but practically we would count on having 7 cycles a year, because of delay by stocking, cleaning etcetera. The stocking density at the end of the growth stage is 500 pieces/m^2 and the tank is 11 m^3; so, we can harvest 5,500 pieces per cycle. The production capacity of 1 pre on-growing tank is 7 x 5,500 pieces = 37,500 pieces.

The total capacity of the pre on-growing to serve the on-growing is 15 x 37,500 pieces = 577,500 post-juveniles. We need 550,000 post-juveniles, so the capacity should be enough.

The mortality rate in the pre on-growing is estimated at approximately 20%. It is important to stock well graded juveniles of 5 g and make sure all the shooters are removed. So, the hatchery should supply 550,000 x 1.2 = 660,000 pieces a year. Each tank in the pre on-growing system is stocked with 6,600 pieces.

What is the reason for having a pre on-growing system?

Some readers might wonder why you need a pre on-growing stage. Why not stock the juveniles from the hatchery directly in the on-growing system?

- By stocking small fish in a big pond, the pond is not utilized efficiently. This way of stocking increases the growth cycle from 150 days to 195 days which reduces the production capacity by almost 25%. The big tanks are an expensive investment and therefore it's important to use them intensively.
- Stocked juveniles of 5 g need grading after 45 days, because there will still be shooters and runts around. If you don't take them out, you will end up with a varied batch of fish, more cannibalism and a few catfish which don't grow but still consume feed. The feed conversion rate will be negatively affected by it.
- In the stage of growing juveniles to post-juveniles from 5 g to 80 g, the fish are still sensitive and can fall sick. Treatment in a small tank, especially a bath treatment, is easier and cheaper compared to a large tank.
- The juveniles from the hatchery have been raised in borehole water and have to be weaned to river water, especially during the rainy season when the river water is muddy. So, the supply of pre on-growing has two sources of water: river water and borehole water.

The hatchery

The capacity of the hatchery has been estimated at 1,000,000 juveniles. For growing table-sized fish, you need to take out only 660,000 juveniles of 5 g. So, the capacity of the hatchery is enough and even allows you to sell some young fish to other farmers.

10.3 CHECK POINTS FOR THE MANAGER

8 points for the manager to check:

1. The stocks of fish in number, age and weight on a weekly basis.
2. The feed given per tank on a daily basis with a weekly summary.
3. The feed stocks on a weekly basis.
4. Your staff clocking in and out.
5. The budget for petty cash made and controlled on a weekly basis.

6. The stocks of diesel for the generator and its service programme on a weekly basis.
7. The harvest and sales and outstanding customer funds.
8. The safety of the environment for staff and customers, which also needs continuous maintenance.

To check all these points, you need to keep records. Below you 'll see the formats for each of them. The formats are filled in for easy understanding. They don't correspond with each other, and are not based on the examples given before in this chapter.

1. Check the stocks of fish in number and in weight on a weekly basis

Every week you should have the following information available for each tank, and also have a summary for all the tanks together.

For each tank:

- Number of fish: preferably adjust for mortality during the week.
- Age of fish: every week you add 7 days.
- Total biomass of fish: add the calculated weight by dividing the feed gift during the week by the expected FCR.
- The individual weight: take the total biomass and divide it by the number of fish.

Example 17. How to calculate the fish size based on the feed gift and FCR?

Week 21:
Tank 1:
Stocked 5,500 pieces
Mortality: 0
Age: 63 days
Biomass: 825 kg
Size: 150 g

Feed gift tank 1 in week 21:
Monday 25 kg, Tuesday 25,5 kg, Wednesday 26 kg, Thursday 26.5 kg, Friday 27.0 kg, Saturday 27.5 kg, Sunday 28.0 kg. Total amount for this week: 185.5 kg with an FCR of 1.25

Week 22:
Tank 1:
Stocked 5,500 pieces
Mortality: 0
Age: 70 days
Biomass: 973 kg (calculation: 185.5/1.25 = 148.4 kg + 825 kg (former biomass) = 973.4 kg)
Size: 177 g (calculation: 973.4 kg/5,500 pieces = 177 g/piece)

You should make the calculations of example 17 for each tank every week. As a manager you can derive a lot of information from them:

- **You will know the total number of fish on the farm.**
- **You will know the average fish size of each pond and when they will be ready for selling.**
- **You will know the total standing stock in kg of the farm.**

These figures shouldn't fluctuate much. If they get lower, it means you are selling more than you are stocking. They also give you the size of the fish by calculation and not by test cropping. This is possible, because the FCR is derived from earlier harvesting results, and the FCR fluctuations are low if the farm is operating constantly and doing well. Test cropping of fish, in my opinion disturbs the fish a lot. It slows down their growth and it does not even yield a clear record of the size of the fish. Also, you are able to predict the number of fish to sell during the coming months, because your target is age 180 days. Any fish that is 150 days old or older can be sold the next month. Never base your sales on the size of the fish but always on age, whether they are big or small. By doing so, you have a clear and stable growth cycle which makes it possible to predict and forecast results.

2. Check the feed gift per tank on a daily basis with a weekly summary

For every tank make a feeding list which should be filled in during the week. It shows the weight in kg and the particle size in mm of the feed used. The progress of feeding already provides an indication of the performance of the fish. Every day there should be an increase of feed (0.5 kg or 1 kg). Since the fish are growing on a daily basis, they should get more feed every day. A reduction of the feed given each day during the week, is a result of problems in the tank, cold weather, disturbance, no feed available or other problems. This should be noticed and checked.

Many farmers love to keep the feed gift per day constant for one week and after one or two weeks they suddenly increase it. They also like to feed a full bag of feed and jump to two bags after some time. It's pure laziness. It does not work like this. These farmers simply don't want to weigh out the feed. Erratic feeding patterns can never give you good growth and good feed conversion rates. I believe that underfeeding the fish is the biggest problem in Nigeria. Farmers are afraid to spend money on feed for their fish, but they lack the understanding that the profit comes from the feed; the more you feed the more profit you make!

The empty feed bags yield valuable information and should be collected and stored. For example, if the farm uses 15,000 kg of feed during the week, you should have 1000 empty feed bags, which comes down to 83 dozen. If bags are missing, it could be a result of bad recording, incorrect weighing out the feed or theft.

Example 18. Daily feed gift per fish tank or pond

Daily feed gift per tank in kg
Size of feed in mm
Farm: Aqua garden

Pond no	Size of Feed (mm)	Monday	Tuesday	Wednesday	Thursday	Friday	Saturday	Sunday	Total (kg)
1	6.0	61	62	63	64	65	66	67	448
2	4.5	40	41	42	42	42	43	44	294
3	4.5	38	39	40	41	42	43	44	287
4	4.5	46	47	48	49	30	31	32	283
5	Empty								
6	Empty								
7	2.0	20	20.5	21	21.5	22	22.5	23	150.5
8	2.0	20	20.5	21	21.5	22	22.5	23	150.5
9	8.0	70	71	72	73	74	75	76	511
10	8.0	80	80	80	80	80	80	80	560
11	6.0	61	62	63	64	65	66	67	448
12	4.5	40	41	41	42	42	43	44	293
Total (kg)		476	484	491	498	484	492	500	3425

3. Check the feed stocks on a weekly basis

Checking the feed stocks can be carried out by taking physical stock every week. In order to do this, each feed size has to be stocked properly and in such a way, that there is easy access for checking every pallet.

During the week there will be changes; stocks will reduce when feed is given to the workers to feed the fish and they will increase when feed is purchased. The feed that comes in, should also be added to the stock. The changes should reflect the consumption of feed during the week. Afterwards, the storekeeper should provide the farm manager with the records.

The records of the feed stocks are important, because they will inform you when you will be out of stock. For example, a farm which feeds around 3,000 kg on a daily basis, should determine the minimum stock level which depend solely on the lead time between ordering and getting the feed and the reliability of the feed company. To be on the safe side for not running out of feed, the minimum stock level before ordering should be the daily consumption x 2 times the lead time.

Example 19. How to calculate minimum stock level of feed?

Suppose the daily feed gift is 3,000 kg
And the lead time is 5 days.
Minimum stock level: 10 (2 x 5) x 3000 kg = 30,000 kg

Example 20. Format for keeping feed stock records

Feed stocks: Aqua garden's weekly physical stock taken at Monday morning before feeding

Feed size (mm)	Previous stock		Present stock	Actual consumption	Consumption according tank report
Product	**kg**	**received kg**	**kg**	**kg**	**kg**
0.5	50	0	30	20	20
0.7	20	100	55	65	60
1	30	150	80	100	100
2	350	250	300	300	300
3	1500	1000	915	1585	1570
4.5	4005	3000	4500	2505	2490
6	3015	6000	7470	1545	1560
Total	**8970**	**10500**	**13350**	**6120**	**6100**

The records in the table show a discrepancy of just 20 kg. According to the physical stock taken, the farm consumed 6,120 kg, but according to the fish farm manager he took only

6,100 kg. The conclusion is that 20 kg is missing, which can be due to mistakes in stock taking, a mistake in weighing out the feed or 20 kg has been stolen. 20 kg is nothing to worry about on a total consumption of 6,000 kg, but if it happens regularly and becomes 100 kg or more, action should be taken.

4. Check your staff clock in and out

If you want to create discipline with your staff, you have to start an attendance list where they can sign in and out and note the time of day when they come in and go out of the farm.

During the day, the staff should remain on the farm or have written permission to go out, given by the farm manager. At big companies they have a time keeper to monitor these things. Normally, he is assisted in doing this by the gate man.

If the farm is small, it is the personal responsibility of the staff to remain at the farm and carry out the work which has to be done. But the moment a farm becomes big, many workers believe that if they are not doing their job, somebody else will do it.

Workers who are at the farm during the night, should by no means go out. Because, if anything happens at the farm, such as robbery, theft or any other criminal offence, the people who have left the farm illegally, are the first suspects!

The attendance list shows the manager the loyalty of each staff member which can be used for decisions about future bonuses. But primarily the list clearly shows the days worked by each worker and their corresponding salaries. As a manager you should keep a clear-cut distinction between you and your staff. You can be friendly to them and interested in their well-being, but you should not become emotionally attached to them, in order to keep a clear unbiased eye on each of them.

Example 21. Format for an attendance list for both day time and night time workers

No	I.D.	Name	Function	Morning	Evening

5. Check and budget for petty cash on a weekly basis

Every week the manager should budget for petty cash needed for the coming week. This money will come in on Monday morning, and by the end of the working day, the money should be counted and balanced. All the money going out should be accounted for! If you take the budget from the sales activities, it will become very blurry and difficult to get clear records of expenses and turnover.

It is best to have a limited number of suppliers which provide you with credit or allow you to pay by bank transfer. The fewer cash transactions the better; if the sausage is close to the dog, the dog will get tempted and eventually eat it, regardless the level of training! A large income of cash attracts criminals and it is very difficult to handle these kinds of situations.

Example 22. Format for a weekly financial report and budget request (cash)

Farm: Aqua garden

			week 9	*week 10*
S/N	*Item description*	*Previous request (₦)*	*Cash expenses (₦)*	*Next request (₦)*
1	*Fuel*	*20000*	*15000*	*21000*
2	*Spare parts*	*0*	*0*	*4000*
3	*Antibiotics*	*16000*	*16000*	*0*
4	*Water test kit*	*4000*	*4000*	*0*
5	*Plumbing materials*	*35000*	*25000*	*7000*
6	*Electrical materials*	*0*	*0*	*8000*
7	*Medical expenses*	*4000*	*3000*	*0*
8	*Stationary, computer*	*11000*	*10000*	*0*
9	*Transport*	*3000*	*3000*	*3000*
10	*Canteen, beverages*	*2000*	*2000*	*2000*
11	*Fish feed*	*15000*	*13000*	*15000*
12	*Others*	*5000*	*0*	*5000*
	Total	***115000***	***91000***	***65000***
	Total available on site		***24000***	
	Total transferred to site			***41000***

6. Check the diesel stocks for the generator and the service programme on a daily basis

In some farms the cost of diesel exceeds the cost of labour! So, it's obvious that the diesel consumption has to be checked and controlled by the management. Diesel is widely used and, once stolen, easy to trade and convert into money.

Depending on the size of the generator and farm, you should have a calibrated tank, so that you can dip on a daily basis to find out how much diesel is in the tank and how much has been consumed by the various machines per day. If the generator is small, you can work with 50-l jerry cans. The moment you fuel the generator tank, you note the volume and check its running hours. What you want to find out is the consumption per hour of the generator. If you know this, you can be aware of changes in the consumption of the generator per hour, and you are also able to predict the consumption for a period of time. Always refill the generator till the brim. Running a generator on a low level of diesel in the tank, is not good for the engine and you also lose control over the consumption of diesel per hour.

When you combine a generator with electricity from the PHCN, you should have an alarm system for when the PHCN-electricity goes off or comes back, in order not to overuse the generator or have unnoticed power outages.

Some farmers keep very low stocks of diesel or petrol and fail to have constant power for pumping water or for running their hatchery; don't be surprised that you don't get good results at your farm. Minimising the cost of energy is actually penny wise and pound foolish!

Servicing the generator, either a petrol or diesel one, is of utmost importance if you don't want any nasty surprises; make a servicing programme and follow it. Servicing is different from fixing the generator when it gets faulty.

The consumption of a generator depends on the load. The table in Figure 30 shows the consumption of diesel per hour in relation to the load and the size of the generator.

Generator kVA	Generator kW	1/4 load l/ hour	1/2 load l/ hour	3/4 load l/ hour	Full load l/ hour
10	8	0.9	1.2	1.7	2.1
12	10	1	1.4	2.1	2.6
15	12	1.3	2.1	2.6	3.2
25	20	2.3	3.4	5	6.1
40	30	5	6.8	9.1	10.9
50	40	0.1	8.7	12.1	14.3
75	60	6.8	11	14.4	18
95	75	7.5	12.9	17.4	23.1
125	100	9.8	15.5	22	28
155	125	11.7	18.98	26.9	34.4
170	135	12.5	20.4	28.8	37.1
190	150	13.6	22.3	31.8	41.3

Figure 30. The diesel consumption per hour in relation to the load and the size of a generator. The figures shown are estimates and variations due to operational differences should be considered. A generator runs efficiently when it is loaded for about 3/4 of its capacity. 1 kW = 1.25 kVA.

Example 23. Format for fuel requisition

Fuel requisition

Farm: Aqua garden Date: 11/3/2019

Filling station: Total filling station opposite Jabi upstairs Time: 10.00 am

Car no: VR- ECK- MAR

Mileages					
Product	Litres	Previous	Present	Difference	km/litre
Diesel					
Petrol	51	86,499	87,005	506	9.9
Engine oil					
Name:					
Driver	attendant	authorized			
Friday	Musa	Wale			

Example 24. Format for fuel requisition for a generator

Fuel Requisition
Site: Aqua garden Date: 11/3/2019
Fuel station: Aqua garden diesel tank time: 10.00 am
Generator: SP-UIT-11

		Hour readings			
Product	Litres	Previous	Present	Difference	Litres/hour
Diesel	230	543	587	44	5.23
Petrol					
Engine oil					
Name:					
Operator	attendant	authorized			
Adama	Monday	Wale			

7. Check the harvest and sales and outstanding customer funds

Cropping and sales are actually two different things. When customers visit the farm, the best way to sell them the fish is at a point of sales. When you sell the fish directly from the pond, all your customers are at your farm at once and become a threat for biosecurity. Your customers carry baskets from one farm to the other and they don't have any idea of biosecurity. The fish in the ponds close to the pond you are selling the fish from, will also be disturbed by all the activity. It is advisable to harvest the fish and carry them to a point of sales close to the gate.

When you harvest the fish, you should weigh them too. Then you can sell them at the point of sales per basket of 25 kg. The people who harvest the fish should not be the same ones who sell them. At the end of the sales day, you compare the harvest results to the sales results. The difference between them should not be more than 3%. The difference arises from not being able to sell exactly 25 kg but 25.3 kg or 25.5 kg. Your customers are very keen to get a little bit more, since the fish lose weight during transportation but they will have to sell the basket for 25 kg to their customers. It is best to use a digital scale and a calibrated weight of 10 or 25 kg to convince your customers that your scale is accurate. See below example 25 showing harvesting and sales results in a month's period.

Example 25. Format for sales and harvest report.

Date	Harvested (kg)	Sales (₦)	Sales (kg)	Difference (kg)	
03/01/2019	5,430	3,909,600	5,270	160	
24/01/2019	6,320	4,518,800	6,190	130	

8. Check safety and maintain a safe environment for staff and customers

As a manager you should check the safety of the farm and maintain a safe environment by following these rules:

- It is an obligation to keep the working environment safe for your staff.
- Operating machinery should be carried out with hand gloves, helmets, ear protection, glasses etcetera.
- Working with chemicals, such as fish medication, should be executed with personal protection.
- Harvesting fish should be carried out with a wet suit, rain boots etcetera.
- Fire extinguishers and the first aid kit should be available and up to date.
- Rats and mice should be eradicated by avoiding spillage of fish feed, and to keep the environment clean. Use rat poison in order to avoid diseases such as Lassa fever.
- Waste removal has to be done regularly.
- The staff should be educated that working safely and keeping the environment clean is every body's responsibility.

11

CATFISH MARKETING

1. INTRODUCTION
2. PRODUCT FORMS
3. PRICING
4. PROMOTION

11.1 INTRODUCTION

Most farmers have no idea how to sell fish when they start farming. They believe that catfish will sell anyway and that there are never enough of them in the market. Yes, globally this is correct; Nigeria still imports a lot of fish, but that doesn't mean that all imported fish can be replaced by farmed Nigerian catfish. Many Nigerians also like to eat seafish, shrimps, tilapia, fish products such as Titus, and other canned fish products. So, to be successful in fish farming requires a plan and an innovative approach to the market.

Wholesalers

The easiest way to sell fish is by contacting the wholesalers or high-volume buyers who buy all your fish at once. If you are producing large quantities of fish, in principle you have to work with buyers who are able to distribute the fish to the markets in Nigeria. These kinds of customers are able to buy all your table-sized fish and transport them as far as a full day's journey. They take responsibility for transport and take care of collecting the money for the fish from the customers, mostly retailers, who sell fish too. The major disadvantage of wholesalers as customers is that the price you get for your catfish is controlled by them. When you are a small-scale farmer, it could be less profitable for you to venture into selling to them, since your cost price is relatively high because you don't farm at the ideal economic scale.

Niche markets

As a small-scale farmer, you should look at markets which are smaller and closer to your farm: these are called niche markets. This way, you 'll become a marketer and in some cases even a retailer, alongside being a fish farmer. It takes time to analyse and develop these niche markets but you have more control over the sales price and it will be higher.

Plan and planning

Marketing requires a plan and planning. How much fish must I sell within a certain time to have a balanced standing stock? The standing stock is the quantity of fish at the farm and it can be derived from the production records. If the standing stock increases, the number of fish you sell, is lower than you produce, which means that your fish will grow bigger and are at the farm longer than necessary. This way more farm space will be occupied and the production will go down. On the other hand, if you sell more than you produce, you will soon be out of production and the fish available for sale, become smaller in size. You will soon have empty ponds and the production will reduce drastically.

Target size

Another inadvisable way of selling is by top cropping. Top cropping means grading the big fish out and leaving the smaller fish in the pond. The advantage of this approach is that you start to sell earlier and initially your cash flow will increase. However, at a later stage, it will be detrimental, because the smaller fish that are left in the pond, will not grow anymore because of the disturbance of the top cropping activity in their pond. Neither do you make optimal use of the pond space. Always sell according to your planning and let the target size be the average size of the fish in the pond.

If the market is strong, you will be tempted to over sell, but you will certainly regret this. Stick to your growth cycle and the size of fish you want to sell. On the other hand, keeping the fish in the pond because the price is not favourable to you, can become very expensive, since the fish have to be maintained in the pond and consume feed, while they don't grow rapidly anymore. Sometimes you have to reduce the price in order to keep the standing stock constant and the production going.

Good and bad prices

In general, the price you get for your catfish, should not be calculated based on one month, but on a production cycle of one year. You will have good and bad prices, but it will balance out. Many farmers are discouraged to buy juveniles when the table-sized fish price is low, which is a big mistake. The time of excess fish is actually the best time to buy juveniles because the market will be strong in six months' time. However, the best approach to marketing is to produce as efficiently as possible and have a constant production at the lowest cost price.

Competition

As already said, catfish are not the only fish eaten in Nigeria. There is a lot of competition from river fish and sea fish. Frozen fish imported from Europe, has been very popular, but it has become expensive since the government introduced import regulations to protect the Nigerian fishing and aquaculture industry. Also, the fishermen's supply varies during the year. Most fish are caught during the dry season from January to April and these supplies certainly compete with the farmed fish.

The number of catfish that are produced varies too; the supply is highest at the beginning of the dry season and lowest when the rains start. Catfish normally sell easily from March to April. The variable supply of catfish is due to the rainfall. Many farmers are able to produce more fish during the rainy season but have problems keeping the fish in the ponds during the dry season and therefore want to sell by any possible means during this period. The middle men take advantage of this situation. The 'rainwater-farmers', who must plan their production in relation to the rainwater availability, have a big disadvantage in comparison to farmers who do not have to rely on this water source.

11.2 PRODUCT FORMS

The catfish industry is dominated by only two products: live catfish and smoked catfish. You rarely see any other product around. This is a logical development since catfish can be easily kept alive with a little water and customers love to buy live fish, especially in Nigeria where the so-called cold chain is not well developed.

Live catfish

The optimum proof of freshness is a fish that is still alive and moving. Roadside catfish sellers know this very well and shake the catfish in order to draw attention to the traffic passing by and to show the fish are still alive!
The popular beer joints have their slogan "Point and kill". The most popular way to eat catfish is in a pepper soup, which definitely also helps to increase the beer consumption. They offer either a complete small catfish or a part of a bigger catfish in the soup. You can choose either head, middle or tail. When the fish is cut in pieces, the retailer can make good money out of one fish. Therefore, the demand for big fish is high, which results in a higher price per kg. The catfish can also be grilled with spices and vegetables wrapped in aluminium foil. This dish comes with chips or any other type of high carb food. Big fish are popular in this type of dish, because people like to share the fish with two or three others.

As a farmer you are unlikely to stand out with live catfish; there is no brand for it and the fish cannot advertise for you. The moment the seller has them in his basket, the fish have become anonymous. This doesn't mean that bad fish sell as well as good fish. If a farmer wants customers to come back and pay a good price, he needs to be aware of this.
With product forms which have a package, the package reveals the source of the product and can advertise for itself.

The following quality aspects should be taken into consideration when selling live fish:

- The gut of the fish must be completely empty, so no feeding 36 hours before selling. If not, the fish will vomit and deposit their waste in the water which will make the water and fish start to smell bad in no time.
- The fish should be healthy, having no wounds on the body or any other abnormalities.
- The fish should look well fed. Lean fish have more bone than meat, which is not desirable for the end-user!
- The female fish should not be full of eggs with a bloated belly. Some people like the eggs, but most people don't.

- The catfish should not have a lot of belly fat, because it is not eaten and therefore a waste. Fat is deposited in the body when the energy level of the feed is very high. The energy will be stored as mesenteric fat and will not be used for growth.
- Fish which are ready to be sold, should be graded by size.

Catfish transport

Marketers buy fish and load their vans and cars with plastic bowls covered with sacks. Each bowl contains 25 kg of catfish and 10 to 15 l of water. The sack is made of a woven material which allows air to pass through. In order to avoid foaming, they put a bit of palm oil in the water by using a 5-ml cap of an ordinary plastic bottle. Planks between the bowls are used to support them. In this way a van can carry over 40 bowls filled with 1000 kg of fish. If the car is being driving, open windows will provide enough ventilation. But, if the car is stationary, the bowls with fish become overheated and the accumulation of carbon dioxide in the water can easily weaken the fish or even kill them. Carbon dioxide gas is heavier than air, so the accumulated gas will remain right above the surface of the water in the bowl. As a result, the fish will suffocate. Retailers will know that the fish have been struggling to stay alive, because when you cook these catfish, their meat is not firm anymore. Also, the end-users who eat them will not enjoy the meat, because its texture has become mushy. That is why the retailers should also consider how to store the catfish in order to maintain good quality. Almost all retailers use bathtubs, but they should not cover them in order to allow ventilation. Change the water once or twice a day, but never use city council water because of the added chlorine.

Smoked catfish

Smoked catfish are used in soups such as the very popular Egusi soup. There are two ways of smoking:

1. Smoked for a short period of time, provides the catfish with its particular smoke taste and preparation, but the fish have not yet been dried. They should preferably be eaten the same day they are bought or they should be kept in the fridge.

2. Smoked for a longer period of time provides the catfish not only with its particular taste and preparation but also with the drying necessary to keep them in good condition without refrigeration. The moisture content of the fish can be as low as 10 - 15% and the fish will have lost a lot of weight. Therefore, most of the smoked fish in the market are sold without knowing the weight but by eye!

Smoking fish can be carried out in an oven or kiln with charcoal, wood, electricity or gas (Photo 39). The oven is usually double sheeted in order to keep the heat inside it. If you as a farmer smoke the fish yourself, you can develop a brand if the fish are properly packaged. Some fishmongers smoke sick or dead catfish, because the smoking process camouflages the freshness of the fish. This is a bad development. Therefore, I advise you to package and brand your catfish, when you smoke them yourself.

Photo 39. An oven or kiln for smoking catfish.

Catfish fillets

Another catfish product which is not yet very popular in Nigeria, is catfish fillets. The catfish industry in the USA, Vietnam and Europe, relies on filleting the fish and packaging them either fresh on ice in insulated Styrofoam boxes or frozen in sealed plastic bags. Value can be added by marinating the fillets with different flavours. Catfish fillets are quite neutral in taste which can be adjusted with herbs to the consumer's preference. By filleting the fish, the product has become consumer-friendly; they just have to fry the fish fillets to eat them. African consumers spend less and less time cooking and are more willing to buy convenience products. With regard to the increasing middle-class population in Nigeria, I believe that boneless fillets could also be distributed and marketed via supermarkets all over the country. Be aware that by filleting the catfish you will lose 55% of the initial weight. The waste from filleting, such as the catfish heads, can be sold to local markets.

11.3 PRICING

Catfish prices vary from selling at cost price to high prices which only a few people are willing to pay. The real price you will get for your fish will be something in between.

Factors influencing the price of your fish:

- Where in the market do you position your product?
- Who are your customers? Are they consumers, pepper soup joints, supermarkets, road side retailers or wholesalers?
- What are your competitors offering?
- Do you have something unique to sell? Or can you differentiate your product or service from other farmers?

Competitive pricing

The easiest way of pricing is to calculate your cost price and add your percentage on top of it. This system works fine when there is not much competition, because you don't have to consider the pricing of your competitors. The most regular way of pricing your product is competitive pricing, which strongly relates to prices offered by your competitors.

Discount pricing

Discount pricing can be useful when you start your business and you want to have a share in the market more quickly. Discount pricing however, is suitable for a period of time but is not limitless. You can also see it as an introductory price. When you are already in the business, a discounted price can be given as an incentive for buying large quantities at once or to very loyal customers. Discounts can also be given if customers pay in advance which can be useful if you have cash flow problems.

In general, different prices for individual buyers will give you a lot of problems. Your customers will not be able to keep their prices secret and you will soon have big arguments. Some farmers think they are smart by keeping the prices the same for everyone, except for a selected group of customers to whom they give away, for example 1 bowl of 25 kg for every 500 kg they buy. Even this cannot work, because the buyers are clever enough to find out the real prices! My advice is to keep the prices the same for everybody except when you have a good reason, such as higher volumes, advanced payments etcetera.

Non-price factors

Pricing the fish differently or pricing them higher because you want to promote your fish on non-price factors, such as healthy fish, grown in clean water, good quality fish feed, does not work in the Nigerian market because, as already stated before, the moment the fish are put in a basket, they become anonymous. People don't believe much in these factors, although a significant group of Nigerian consumers believe that catfish from the wild taste even sweeter than farmed catfish.

Knowledge of the market

Fish farmers generally do not collaborate. They will never tell their colleague farmer the real sales prices. If they do, they will most likely tell you a higher price than the one they sell at. You start wondering why you cannot sell at these prices your competitor can! It's obvious, it is because he is selling for a lower price than the price he tells you.

Most farmers have limited knowledge of the market situation. Is the market flooded with fish or is there a shortage? They don't know. The wholesalers take advantage of this situation and quickly buy from farmers who are isolated from the real market situation, telling them the market is dull and fish is barely moving, while in fact the opposite is true! By the time these farmers find out the truth, they have already sold most of their tablesized stocks. Figure 31 shows the prices for catfish in 2018 from a farm in Abuja.

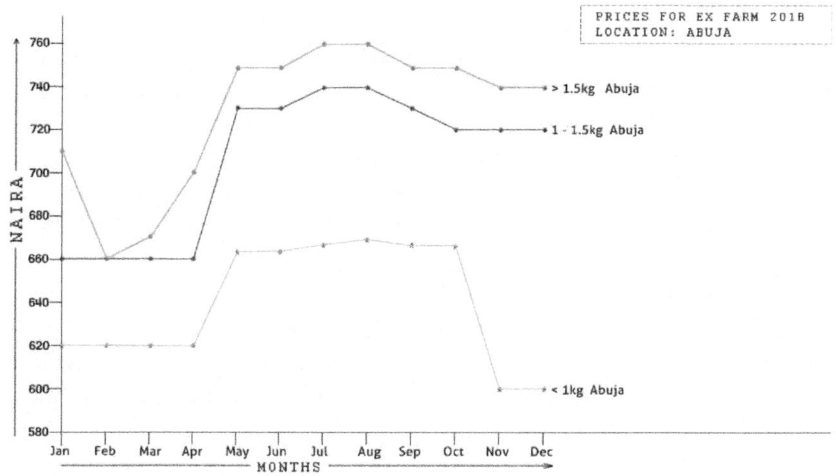

Figure 31. Prices for catfish from a farm in Abuja in 2018.

Price trend

As you can see in Figure 31, catfish prices fluctuate during the year depending on availability and demand. Apart from this, inflation also plays apart; normally the price should increase because of inflation. During the rainy season, the price of fish tends to rise. You will also see this trend in other years. In addition, you can see that fish smaller than 1 kg are always lower priced than bigger fish. That is partly caused by the fact that many farmers sell small fish because they are not able to produce bigger ones. The most interesting part of this chart is the price trend shown, not the absolute figures. In Oyo State for instance, the prices shown in the chart will not be possible for farmers, but in the North of Nigeria and in Port Harcourt, they are not only possible but can be even higher.

11.4 PROMOTION

The moment you have a product and a price, you start your promotion. Promotion is usually personal and for your product only, but I want to emphasize that promotion of fish products should also be carried out generally. What does this imply?

Fish products should be positioned in the market as healthy and tasty. The health aspect of fish for older people and also for children should be emphasized via magazines, articles, brochures etcetera. This can be done by umbrella organisations, such as CAFAN and organisations in the fishing industry. The consumer makes a choice to eat fish or any other meat product, but will be more likely to decide to buy fish if it is widely promoted. When the demand for fish generally increases as a result of this, it will also increase the demand for catfish. The problem with this kind of promotion however is that everybody involved in the fishing sector should contribute, which is almost impossible to do in the Nigerian society. Most Nigerians want a free ride. That's why general promotion happens only by the grace of reporters, books, medical staff, teachers etcetera. It is not coordinated at all.

Advertise channels

The most common way to promote your own product is by personal communication. This can be highly effective, but word of mouth promotion is slow going. Also be aware that positive word of mouth promotion comes by feet, but negative word of mouth promotion leaves on horseback!

Other advertising channels include radio, newspapers, TV, magazines, handbills, flyers and posters. The promotional message must be clear, to the point and focused.
When you visit the bigger fish markets in town, you will recognize your potential customers straight away. In some areas they specialise in smoked fish, and you will notice your potential customers there as well. You can hand out flyers to them or try to convince them to visit your farm to show them your fish.

Other places to sell fish are retail shops, road side markets, locations close to big offices and fairs and festivals. It just depends on how inventive you are and how much time you want to spend marketing your fish.

Delivery of fish

Sales which include delivery can be very tricky, because the moment you arrive at the customer's place, he is the one who is in charge; you want to sell your perishable product, but if he does not mean well, he can take advantage of this. Delivery of fish can only be done in mutual understanding with the buyer and when you are sure he is a fair business man.

Also, be aware that a sale has only been carried out successfully after payment has been fulfilled!

12

ECONOMICS OF CATFISH FARMING

1. DIRECT COSTS
2. INDIRECT COSTS
3. CALCULATION OF BREAK-EVEN POINT
4. FARM SIZE VERSUS PROFITABILITY

12.1 DIRECT COSTS

Expenses for fish feed and juveniles are direct costs, because they are directly related to the fish production.

As already stated in chapter 7, fish feed represents the biggest part of the costs to produce catfish. Depending on the system and efficiency of the farm, it can increase to 75% of the cost price. The higher the percentage, the better it is, taken in consideration that the feeding and the feed quality are okay. At large farms with a high production the percentage is high, while at small farms the percentage is low. The overheads, which are indirect costs, are relatively higher for smaller farms than for larger farms. That is the advantage of scaling up production. For example, a farm producing 20 tons of fish a year is able to increase the production to 50 tons a year, while the costs for manpower, security, light etcetera will not increase 2.5 times. These costs will increase less than that.

Next to feed, expenses for juveniles are the other direct cost of a fish farm. The number of juveniles needed to reach a certain production, is dependent on the selling size of the fish (Figure 32).

Selling size of the fish (g)	Number of juveniles/kg	Mortality 20%	Total	Cost (₦)
500	2	0.5	2.5	50
1000	1	0.25	1.25	25
1500	0.67	0.17	0.84	17
2000	0. 5	0.125	0.625	12.5

Figure 32. Cost of juveniles to produce 1 kg of fish, taking the selling size in consideration. Cost of juveniles: ₦ 20/piece. Mortality rate from juvenile to table size: 20% due to cannibalism, diseases, birds etcetera.

When you buy good quality juveniles and grow them to table size, you will still lose 20% due to mortality, mainly at the beginning of the growth cycle. So, if you sell them at 1 kg average, you 'll need to purchase 1.25 juvenile for each kg. If the cost price for juveniles is ₦ 20/piece, your cost for juveniles to produce 1 kg fish, will be ₦ 20 x 1.25 = ₦ 25/kg. When you produce fish for smoking and selling them at 500 g, you will need 2.5 juvenile for each kg of smoked fish produced. The cost will be 2.5 x ₦ 20 = ₦ 50/kg.

As you can see, the size you sell your fish at, matters a lot when it comes to the number of juveniles you need and the cost of juveniles per kg of table-sized fish. To make it even more complicated, the feed conversion rate also changes with the selling size. Big fish have a higher FCR than smaller fish. Another factor which varies, is the selling price for the different sizes of fish. Normally bigger fish will fetch a higher price per kg. The FCR varies from 1 for 500 g fish, to 1.38 for 2000 g fish (Figure 33).

Selling Size of Fish (g)	Cost of Juveniles (₦)	FCR	Feed Cost (₦)	Feed cost to produce 1 kg	Cost price to produce 1 kg (₦)	Sales price (₦)	Gross margin (₦)
500	50	1	380	380	430	590	160
1000	25	1.15	370	426	451	650	199
1500	17	1.25	370	463	480	720	240
2000	12.5	1.38	360	497	510	740	230

Figure 33. Relationship between the size of the fish and the gross margin. The figures have been obtained from a farm in Abuja at the end of 2018.

From the figures of a farm in Abuja, shown in Figure 33, you can conclude form the gross margin that the selling size of 1.5 kg is most profitable. However, there are other factors which have to be taken into consideration; for instance, the market demand for this fish size. Actually, the farm in Abuja is targeting at 1.5 kg fish because the competition with local small farmers is less for bigger fish sizes than for sizes of 1 kg or less. Many farmers don't have the financial capacity to grow bigger sized fish.

12.2 INDIRECT COSTS

Indirect costs are costs that are not directly accountable for the production of fish. These costs are less variable and can even be fixed. Such costs are:

- Administration
- Security
- Energy
- Marketing
- Labour
- Depreciation
- Interest
- Maintenance
- Miscellaneous

In principle, the direct costs for all the farms are the same, while the indirect costs vary a lot from farm to farm. The location, the facilities and the capacity of the farm matter a lot.

For instance, do you have a steady access to the supply of PHCN or does your farm run on a generator most of the time? Is the accessibility of your farm good enough when it comes to marketing, transport and even the sales price of the fish? Has the farm been built with good materials or has it been constructed in an inferior way which results in the need for a lot of maintenance? Do you have good staff and a high level of integrity on the farm? Do you fund your farm with a bank loan or with your own money?

12.3 CALCULATION OF BREAK-EVEN POINT

Each farmer should know the minimum production capacity of his farm needed to pay back its running costs. The remaining profit is for refunding the investment and paying the interest if applicable. It is very important to know the break-even point. The following example and Figure 34 show how this is calculated.

Example 26. How to calculate the break-even point?

The farm has a capacity of 100 tons per annum and the fish are sold at the average price of ₦ 720/kg. The direct costs (see Figure 33) are ₦ 480/kg. Gross margin = 720 – 480 = ₦ 240/kg.

The following indirect costs per annum are applicable per annum and estimated in ₦ currency:

- Administration and labour; 4 people - 2,000,000
- Security; 2 people - 1,000,000
- Energy - 2,600,000
- Marketing - 1,000,000
- Interest (financed through own resources)
- Maintenance - 500,000
- Miscellaneous - 500,000

Total: ₦ 7,600,000.

The break-even point represents the amount of production that is necessary to pay for the running cost of 7,600,000 : 7,600,000/240 ₦/kg = 31,666 kg.

Achieved production	Gross profit (₦)	Costs of running the farm (₦)	Net profit (₦)
10 ton/annum	2,400,000	7,600,000	Loss 5,200,000
31.7 ton/annum	7,600,000	7600,000	No gain no loss
50 ton/annum	1,200,000	7600000	Profit
			6,400,000
75 ton/annum	18,000,000	7,600,000	Profit
			10,400,000
100 ton/annum	24,000,000	,7600,000	Profit
			16,400,000

Figure 34. Calculation of profits in relationship to the amount of production.

When you put the figures in an Excel sheet you can play with them by changing the selling price, the FCR, the feed price or the labour cost. You will immediately see the changes in gross profit and net profit of the farm. Also, the break-even point will change.

The example shown in Figure 34 comes from a financially healthy farm; even when the farm produces only 31% of its production capacity, it will not yield a negative cash flow which means that the farm can be maintained. The problems start when a farm has to produce close to the maximum capacity to break-even, because then the net margin will be very low.

Prices for catfish during the year

The prices for catfish fluctuate during the year depending on availability and demand. Apart from this, inflation plays a part. Normally the price should increase because of it. During the rainy season, the price of fish tends to increase. This is an annual trend. You can derive from Figure 31 in chapter 11, that fish smaller than 1 kg are always sold at a lower price. This is partly due to the fact that many farmers sell small fish and few farmers are able to produce bigger fish.

12.4 FARM SIZE VERSUS PROFITABILITY

For several reasons the profitability of farms varies not only a lot between farms of the same size but also between small, medium and large farms. First let us categorise the sizes of farms.

Small farms: less than 10,000 kg per annum

Small farms are usually farms producing less than 10,000 kg per annum and are mostly backyard farms or cluster farms. The definition of backyard farms has already been given in chapter 4, but in short it's a farm which combines a household with a little fish farm in the backyard. A cluster farm is a group of farms situated in a water logged area run by many individuals who share the responsibility for security, roads, market etcetera.

Medium sized farms: less than 100,000 kg per annum

Medium sized farms are usually situated on the outskirts of a big town where the owner lives.

Large farms: over 100,000 kg per annum

Large farms are usually located in areas where there is enough space and water to facilitate a massive production of fish. The owner can live anywhere in Nigeria or even outside Nigeria.

Statistics about the number of farms in Nigeria and their productions are either not available or not reliable, but I believe that Nigeria has over 100,000 small-scale farms while there are up to 20,000 medium sized and up to 300 large farms.

Differences in profitability

The capacity of the farm and its actual production are in most cases not in line with each other. The medium sized farms especially claim large production capacities but some of them produce even less than 10% of their capacity and they should in fact be categorised as small-scale farms, but without the benefits of a small farm.

The small sized and the medium sized farms especially show very big differences in profitability within their categories. At large farms the way of farming has been harmonised to the best farming practice and therefore the range in profitability between these farms is not as much as in the other categories. Large farms are large for a reason: they do well and have the correct infrastructure and a good market.

Medium sized farms face the biggest problems in my opinion. The ambition of medium sized farms is bigger than what they can handle. Most of the time they face lack of good management and security, lack of finance and lack of good infrastructure, such as the availability of water. They are too small to hire good staff and are not able to build good infrastructure. Finally, the set-up of a medium sized farm is too big for the farmer to handle. Most of the time the locations where the farms are situated, are too far away for proper supervision and therefore the farms are only checked at the weekend, in the evening or by phone.

The smaller farm's set up is not ambitious but smart; they don't pretend to make a lot of money but at least the majority of them make a profit. The quantities of fish produced can even be sold to retailers and these farms will get a better average price than the medium sized farmers. Drop outs are usually found in the category of the medium sized farms.

The fact that Nigeria is the biggest fish farming nation on the continent, apart from Egypt, is not because we farm any better, it is because of the mentality of Nigerians; Nigerians don't shy away from risks, have entrepreneurial skills and don't give up easily. Soldier go, soldier come, barrack remains the same!

13

INTEGRATED FISH FARMING

1. INTRODUCTION
2. 3 METHODS OF INTEGRATED FISH FARMING
3. CATFISH AND CHICKEN SLAUGHTER WASTE
4. USING FISH EFFLUENT FOR IRRIGATION PURPOSES
5. HOW FISH EFFLUENT CAN BE USED IN NIGERIA
6. COMBINING RICE AND CATFISH FARMING

13.1 INTRODUCTION

A farm in the middle of Ibadan, used recirculation systems and had a big settling tank for the waste water in order to keep the gutters and the public drainage system free from its solids. Each year the sludge from this tank was used for the polo fields as a preparation for the polo tournament in January. A big tanker was used to transport the sludge and distribute it over the polo field. However, the horses were so happy with the abundance of grass on the field that they refused to run but instead enjoyed the grazing!!

Integrated fish farming is a system of fish production in combination with other agricultural/livestock farming operations close to the fish pond. The farming sub-systems, for instance fish, crop and livestock, are linked to each other in such a way that the by-products and wastes from one sub-system become valuable inputs for another sub-system and thus ensures total utilization of the land and water resources of the farm, resulting in a maximum and diversified farm output on minimum overheads and labour costs.

13.2 3 METHODS OF INTEGRATED FISH FARMING

1. The waste (water and nutrients) of the fish are collected and become a source of nutrients and water for rearing crops.
2. The waste from live stocks or crops is given to the fish pond where it becomes a direct feed source, for instance chicken intestines and faeces, or it becomes a fertiliser to enhance the growth of algae and zooplankton which will also be a source of fish feed.
3. A combination of the first 2 options.

In Nigeria integrated fish farming is rare, but in Asia, for the small-scale farmer especially, it's a beneficial way of farming and therefore popular. Farmers use the manure of live-stock, such as cattle, chickens or pigs, to fertilise the water; in some cases, the pen of the animals is situated above the fish ponds and the droppings fall directly into the pond. The ponds are also relatively big: 0.5 to 1.0 ha. The fish species used in this type of farming are tilapia and several species of carp, either in monoculture or in polyculture. These fish directly consume a little part of the manure but in general the natural production of the fish pond is increased because the manure fertilises the water. The natural food for the fish is algae, zooplankton and benthos.

In Nigeria this type of integrated farming is not popular, because of the fish species we grow. Catfish do not benefit directly from the enhanced natural production as other fish species do, because they are carnivorous and not omnivorous or herbivorous, such as carp and tilapia. Fish production with integrated farming is rather extensive and, in most cases, not more than 5000 kg per hectare per year. In Nigeria land is too expensive to use it extensively. All farms use complete feed or at least additional feed to grow catfish in densities which allow a production of at least 50,000 to 100,000 kg per ha per year.

13.3 CATFISH AND CHICKEN SLAUGHTER WASTE

Catfish do well on chicken slaughter waste, such as fat, blood, intestines and organs that are not eaten by humans, and aborted eggs. This waste can be put into the pond directly or it can be processed into a moist pellet with other ingredients and then fed to the catfish. A farm in Ibadan has been very successful in using chicken slaughter waste directly for the catfish. Their brand is known for extremely big fish! Another farm together with a slaughter house in Lagos, has been very successful with converting chicken waste into a moist pellet for catfish, and using a recirculation system to grow their catfish. They already started this collaboration by the year 2000. These companies have their own chicken farms, slaughter houses, freezers and machines and in both cases the slaughter waste is processed close to the fish farm. Chicken waste appears to be a very good feed for catfish. This feed can also be given in combination with ordinary pellets. However, for catfish below 500 g it is not advisable to feed raw chicken slaughter waste.

It is obvious that the water pollution at the farm will be higher with this kind of feed than with a good quality pellet, so chicken waste can only be used if enough water is available to flush the ponds.

If you are a farmer without a slaughter house and you still intend to use slaughter waste, this could be difficult. Logistics, availability and competition for the raw materials can make it unprofitable to use these products, apart from the effort needed to keep them fresh till feeding time!

13.4 USING FISH EFFLUENT FOR IRRIGATION PURPOSES

In principle the waste water of fish ponds is good for fertilising and watering crops, but it has its restrictions. First of all, the nutrients in the water are available in very low concentrations, so your crops fed only with fish effluent, will not get enough fertiliser.

Researchers in Sagana Institute in Kenya found out that fish water effluent can contribute at most 20% of the nutrients required. After all, you cannot flood the crops in order to push the nutrients out.

To supply the water to the crops or the orchard is another hassle. Normally, a good method is drip irrigation. A small polyethylene hose with small holes 50 cm apart is supplied with effluent under a pressure of only 0.5 bar and dispenses the water in small quantities to the crops efficiently. Also, there will not be much evaporation. The problem with fish farm effluent though, is that the water has been contaminated with a lot of bacteria, so called 'living water'. They cause the holes of the drip hose to clog very easily, quickly resulting in a bad water distribution. Therefore, effluents from fish farms can be used only in an open irrigation system with canals or water sprayers with big holes and lots of pressure.

Another restriction of fish farm effluent is the fact that there is no need for this water during the rainy season.

13.5 HOW FISH EFFLUENT CAN BE USED IN NIGERIA

The fish effluent contains suspended solids, such as faeces, clusters of bacteria and dead algae, which settle rather easily. So, it is advisable to have a settling pond for all the drainage water of the farm. Let the overflow eventually return to the drainage or river, because it will be much less contaminated. The sludge can be pumped out of the settling tank and used to improve the amount of organic matter in the soil and increase its nutrient levels. In many parts of Nigeria, the level of organic matter in the soil is low, because it contains mostly sand. When the sludge is added to the soil, it will become more fertile.

13.6 COMBINING RICE AND CATFISH FARMING

Rice farming has become increasingly popular in Nigeria since the government has promoted rice culture and banned the import of rice. This offers the opportunity to combine rice and fish culture; this way not only carbohydrates are produced, but also the so much needed proteins. This type of farming is very popular in Asia and we can learn from them. Already many trials are conducted in Nigeria. In most cases the combination of rice and fish is more profitable than rice or fish separately.

Fish culture in rice fields can be done in 2 ways:

1. Simultaneous culture: both at the same time
2. Rotation culture: one after the other

Simultaneous culture

Simultaneous culture of rice and fish can best be carried out in rice fields of 0.1 ha divided into 4 subplots of 10 x 25 m. Between each plot there is a ditch 0.75 m wide and 0.5 m deep. Around the plots there are dykes 0.3 m wide and 0.3 m high. Each field has an inlet and outlet both with screens, in order to keep the water at the right level and avoid fish coming in or going out. The ditches are shelters for the fish where they can hide when they are not foraging between the rice plants. When it is time to harvest, the water level is reduced so that the fish can only remain in the ditches and are relatively easy to catch.

Many fish species can be cultured in rice fields. They should be specifically tolerant to high water temperatures (35 °C), low oxygen levels and high turbidity. In Asia the most common species to be cultured in combination with rice are Asian carp, rohu, common carp, milkfish and catfish species. The African catfish can also be cultured in rice fields, since it tolerates these conditions.

Procedure

Five days after the rice seedlings have been transplanted from the nursery site to the pond, catfish juveniles are stocked at a ratio of 4000 per ha and are given supplemental feeding. The fish can grow during the entire period the rice is growing. Approximately 120 days are needed to grow rice from seedling to harvest. In this period the fish can grow up to 500 g (smoking size); depending on the quantity and quality of the supplements fed, the fish might be smaller or bigger.

Advantages of simultaneous culture:

- The fish increase the rice yield by 5 to 15%, mainly due to indirect organic fertilisation through the fish excrements. If stocked with tilapia, the tilapia will eat algae which otherwise would compete with the rice for nutrients.
- Insect pests of rice, like stem borers, are controlled by fish that feed on them, which are in particular catfish.
- Fish also feed on the mosquito larvae which are intermediate hosts for malaria and other waterborne diseases. This way they reduce the risk for human beings becoming infected by one of these diseases.

The major limitations when cultivating rice and fish simultaneously, are the use of agrochemicals to control weeds and insects. Most of those agrochemicals are toxic for fish.

Rotational culture

With the rotational culture of rice and fish, after the pond has been used for fish culture, it is converted into a rice paddy. The main advantage is that during the culture of rice you can use the herbicides and pesticides needed to control weeds and pests. And, during the culture of fish, you can have a higher water level which is more favourable for the catfish. The sludge at the bottom of the pond is an excellent fertilizer for rice.

APPENDICES

APPENDIX 1

Fish farm growth stages of catfish from fertilization to selling size

1. Fertilised egg: the egg after successful fertilization.
2. Yolk sac-fry: the larval stage of fish which has just hatched and feeds on its yolk sac for approximately 2 days.
3. Free swimming fry: fry that swim up in search of food. This a one day's growth stage, also called 'swim up-fry'.
4. Advanced fry: the period after the free-swimming fry up to 21 days.
5. Fingerling: a fry that starts to use its organ to breath oxygen from the air. The moment fry are able to do this, they are called fingerlings. This growth stage lasts until they are 6 weeks old and weigh less than 1 g.
6. Juvenile: a catfish from 1 g to 10 g
7. Post-juvenile or jumbo: a catfish from 10 to 50 g.
8. Grow-out: a catfish from 50 g to selling size.

APPENDIX 2

Cash flow forecast for a 100 tons per annum production capacity catfish farm selling 12 times a year

Note. Interest and depreciation are not included.

Starting up with one tank and stocking another one the next month, implies that you have lower costs for energy and staff at the beginning, and that the costs will gradually increase until your fish are ready to harvest in July.

√ Yearly production: 100 tons. √ Selling every month: 8300 kg. √ Juveniles: 8000 per month, ₦ 25/piece. √ Growth cycle from juvenile to 1.5 kg: 6 months. √ Sales price: ₦ 720/kg. √ Average feed price: ₦ 375/kg.

Costs/Month	January	February	March	April	May	June	July	August	September	October	November	December
Staff incl. admin	100000	100000	120000	150000	170000	170000	170000	170000	170000	170000	170000	170000
Security	80000	80000	80000	80000	80000	80000	80000	80000	80000	80000	80000	80000
Energy	50000	80000	110000	140000	180000	220000	220000	220000	220000	220000	220000	220000
Marketing, sales	nil	nil	nil	nil	nil	80000	80000	80000	80000	80000	80000	80000
Maintenance	40000	40000	40000	40000	40000	40000	40000	40000	40000	40000	40000	40000
Unforeseen	40000	40000	40000	40000	40000	40000	40000	40000	40000	40000	40000	40000
Feed	300000	800000	1500000	2250000	3000000	3500000	3500000	3500000	3500000	3500000	3500000	3500000
Juveniles	200000	200000	200000	200000	200000	200000	200000	200000	200000	200000	200000	200000
Total expenses (₦)	810000	1340000	2090000	2900000	3710000	4330000	4330000	4330000	4330000	4330000	4330000	4330000
Sales (₦)							6000000	6000000	6000000	6000000	6000000	6000000
Cash flow (₦)	-810000	-2150000	-4240000	-7140000	-10850000	-15180000	-13510000	-11840000	-10170000	-8500000	-6830000	-5160000

Maximum working capital needed till first sales: ₦ 15,000,000. This capital is also called the value of the standing stock, which is between 20 - 25 tons. After 15 months the standing stock has been paid and you will have a positive cash flow of ₦ 1,700,000 from then onwards.

APPENDIX 3

Cash flow forecast for a 100 tons per annum production capacity catfish farm selling 2 times a year

This appendix shows you in facts and figures that stocking all the fish at once is not desirable, since you have a very high demand before selling. Apart from this, there are many other reasons not to stock all the fish at the same time:

- Selling all the fish in one month will be difficult and you cannot actually build a customer base this way.
- When things go wrong at the farm; all your fish are at stake! Your standing stock at the point of selling is 2 times higher than when you sell the fish on a monthly basis.
- Your infrastructure should be able to accommodate all these fish at the same time. The consumption of water and power is at least two times higher than when you are selling on a monthly basis at the point of delivery!

DON'T BE GREEDY; STOCK AT REGULAR INTERVALS AND CONSEQUENTLY ENJOY REGULAR SALES!

See the table on the next page.

Note. Interest and depreciation are not included.
Marketing costs are applicable when you are getting the fish ready for sales in June and July. After that period sales activity and associated costs are not included for 6 months. The juveniles are bought in bulk, so you can negotiate a better price: ₦ 20/piece. The other costs have an immediate effect throughout the cycle since you are stocking all the ponds at once.

√ Growth cycle from juvenile to 1.5 kg: 6 months. √ Sales price: ₦ 720/kg. √ Average feed price: ₦ 375/kg

Costs\Month	January	February	March	April	May	June	July	August	September	October	November	December	January
Farm staff incl. admin	170000	170000	170000	170000	170000	170000	170000	170000	170000	170000	170000	170000	170000
Security	80000	80000	80000	80000	80000	80000	80000	80000	80000	80000	80000	80000	80000
Energy	220000	220000	220000	220000	220000	220000	220000	220000	220000	220000	220000	220000	220000
Marketing, sales	nil	nil	nil	nil	nil	80000	80000	nil	nil	nil	nil	80000	80000
Maintenance	40000	40000	40000	40000	40000	40000	40000	40000	40000	40000	40000	40000	40000
Unforeseen	40000	40000	40000	40000	40000	40000	40000	40000	40000	40000	40000	40000	40000
Feed	1000000	2250000	3750000	4500000	5250000	6500000	1000000	2250000	3750000	4500000	5250000	6500000	1000000
Juveniles	960000	nil	nil	nil	nil	nil	960000	nil	nil	nil	nil	nil	960000
Total expenses (₦)	**2510000**	**2800000**	**4300000**	**5050000**	**5800000**	**7130000**	**2590000**	**2800000**	**4300000**	**5050000**	**5800000**	**7130000**	**2590000**
Sales (₦)							**36000000**						**36000000**
Cash flow (₦)	-2510000	-5310000	-9610000	-14660000	-20460000	-27590000	5820000	3020000	-1280000	-6330000	-12130000	-19260000	14150000

Maximum working capital needed till first sales is ₦ 27,600,000. After sales in July you start all over again and the next sales is in January the next year. You will have a positive cash flow of ₦ 14,000,000 but no fish in the farm again. When you stock again in January, the cash flow will be negative again from May till July, when you sell for the third time.

APPENDIX 4

Fish farm from plan to completion in 12 photos

Photo 1A. A design for a grow-out facility for catfish with a storage tank facility at the back; flow-through system.

Photo 2A. Foundation and reinforcements stage 1.

Photo 3A. Foundation and reinforcements stage 2.

Photo4A. Foundation and reinforcements stage 3.

Photo 5A. German floor with hardcore.

Photo 6A. Casting the floor.

Photo 7A. Laying the blocks of the tanks.

Photo 8A. Laying the blocks of the walls and casting the floor of the second line.

Photo 9A. Plastering the walls of the tanks stage 1.

Photo 10A. Plastering the walls of the tanks stage 2.

Photo 11A. Water supply line from the storage tank to the fish tank.

Photo 12A. A view of the gutter and the drainage.

APPENDIX 5

Bill of quantities (BOQ) of a concrete pond and a block pond of 10 x 5 x 1.3 m (length x width x height)

Sample BOQ - Concrete Pond

S/No	Description	Quantity	Unit	Rate (₦)	Amount (₦)
A	**EXCAVATION & EARTHWORK**				
	Clearing and preparation of the site including removal of shrubs and top soil (13.59 x 7.46)m up to a depth of 150 mm and disposed off as directed.	101.38	m^2	200.00	20,276.28
	Mass excavation for Raft foundation(13.59 x 7.46)m not exceeding 0.9 m deep and removal of surplus as directed.	91.24	m^3	500.00	45,621.63
	Backfilling Filling around strip foundations with selected excavated materials: well rammed, compacted and consolidated	31.84	m^3	500.00	15,921.00
	Hardcore Filling				
	230 mm thick approved hardcore filling spread, levelled, well rammed and consolidated to receive oversite concrete	13.04	m^3	4,500.00	58,694.85
	Level and blind surface of harcore with 50 mm quarry dust to receive oversite concrete	4.35	m^3	2,000.00	8,695.53
	Total Excavation & Earthwork				**149,209.29**
B	**CONCRETE WORK**				
	In Situ Concrete				
	Supplying and placing of concrete (1:2:4) as specified to Oversite Concrete 200 mm thick	13.20	m^3	33,000.00	435,600.00
	Supplying and placing of concrete (1:2:4) as specified to Wall 200 mm thick	10.50	m^3	33,000.00	346,632.00
	Formwork				
	Marine board (2.4 x 1.2)m	37.00	No.	11,500.00	425,500.00
	2" x 3" Timber	150.00	No.	700.00	105,000.00
	3" Nail	2.00	Bag	9,000.00	18,000.00
	Total Concrete Work				**1,330,732.00**

C	PLASTERING & SCREEDING				
	<u>Plastering</u>				
	Wall plastering in cement, sand & Stone Dust (1:2:3) mortar n.e 40 mm thick	104.39	m^2	1,500.00	156,585.00
	<u>Screeding</u>				
	Floor Screeding in cement, sand & Stone Dust (1:2:3) mortar n.e 50 mm thick	56.71	m^2	1,500.00	85,065.00
	Total Plastering & Screeding				**241,650.00**
D	REINFORCEMENT				
	High yield steel reinforcement (to BS 4449 and characteristic strength fy = 460N/mm^2)				
	High yield reinforcement bars (Y2)	2.65	ton	272,000.00	720,800.00
	Total Reinforcement				720,800.00
	SUMMARY				
A	**EXCAVATION AND EARTHWORK**				**149,209.29**
B	**CONCRETE WORK**				**1,330,732.00**
C	**PLASTERING & SCREEDING**				**241,650.00**
D	**REINFORCEMENT**				**720,800.00**
	TOTAL				**2,442,391.29**
	ADD 5% CONTINGENCY				**122,119.56**
	ADD 30% CONTRACTOR'S FEE				**732,717.39**
	GRAND TOTAL				**3,297,228.25**

Sample BOQ - Block Pond

S/No	Description	Quantity	Unit	Rate (₦)	Amount (₦)
A	**EXCAVATION & EARTHWORK**				
	Clearing and preparation of the site including removal of shrubs and top soil (13.59 x 7.46)m up to a depth of 150 mm and disposed off as directed.	101.38	m^2	200.00	20,276.28
	Mass excavation for raft foundation(13.59 x 7.46)m not exceeding 0.9 m deep and removal of surplus as directed.	91.24	m	500.00	45,621.63
	Backfilling				
	Filling around strip foundations with selected excavated materials: well rammed, compacted and consolidated	31.84	m^3	500.00	15,921.00
	Hardcore Filling				
	230 mm thick approved hardcore filling spread, levelled, well rammed and consolidated to receive oversite concrete	13.04	m^3	4,500.00	58,694.85
	Level and blind surface of harcore with 50 mm quarry dust to receive oversite concrete	4.35	m^3	2,000.00	8,695.53
	Total Excavation & Earthwork				**149,209.29**
B	**CONCRETE WORK**				
	In Situ Concrete				
	Supplying and placing of concrete (1:2:4) as specified to Oversite Concrete 200 mm thick	13.20	m^3	33,000.00	435,600.00
	Supplying and placing of concrete (1:2:4) as specified to all columns (230 x 230)mm	0.55	m^3	33,000.00	18,155.28
	Supplying and placing of concrete (1:2:4) as specified to all beams (230 x 230)mm	2.30	m^3	33,000.00	75,748.20
	Block hole filling	2.03	m^3	33,000.00	67,095.60
	Formwork				
	Using Fascia Board (25 x 300)mm				
	Columns	16.00	No.	1,400.00	22,400.00
	Sides of beams 230 mm high	28.00	No.	1,400.00	39,200.00
	Sides of oversite concrete 200 mm high	19.00	No.	1,400.00	26,600.00
	2" x 3" Timber	50.00	NO.	450.00	22,500.00
	3" Nail	1.00	Bag	7,000.00	7,000.00
	Total Concrete Work				**714,299.08**

C	**BLOCKWORK**				
	230mm thick hollow concrete block in cement, sand & Stone Dust (1:2:3) mortar filled with concrete	53.10	m^2	3,700.00	196,462.60
	Total Blockwork				196,462.60
D	**PLASTERING & SCREEDING**				
	Plastering				
	Wall plastering in cement, sand & Stone Dust (1:2:3) mortar n.e 40 mm thick	104.39	m^2	1,500.00	156,585.00
	Screeding				
	Floor Screeding in cement, sand & Stone Dust (1:2:3) mortar n.e 50 mm thick	56.71	m^2	1,500.00	85,065.00
	Total Plastering & Screeding				241,650.00
E	**REINFORCEMENT**				
	High yield steel reinforcement (to BS 4449 and characteristic strength fy = 460N/mm^2)				
	High yield reinforcement bars (Y8)	0.11	ton	309,100.00	34,001.00
	High yield reinforcement bars (Y12)	1.63	ton	272,000.00	443,360.00
	Total Reinforcement				477,361.00
	SUMMARY				
A	**EXCAVATION AND EARTHWORK**				**149,209.29**
B	**CONCRETE WORK**				**714,299.08**
C	**BLOCKWORK**				**196,462.60**
D	**PLASTERING & SCREEDING**				**241,650.00**
E	**REINFORCEMENT**				**477,361.00**
	TOTAL				**1,778,981.97**
	ADD 5% CONTINGENCY				**88,949.10**
	ADD 30% CONTRACTOR'S FEE				**533,694.59**
	GRAND TOTAL				**2,401,625.66**

APPENDIX 6

Examples of ranges in feed sizes and their composition

Starter feeds are usually high in proteins with high levels of fish meal and fish oil. Feed conversion rates below 1 are obtainable and can even be as low as 0.6.

Sizes (mm)	Protein (%)	Fat (%)	Crude fibre (%)	Ash (%)	Crumbles
0.2 - 0.3	56	15	0.1	12	sinking
0.3 - 0.5	56	15	0.1	12	sinking
0.5 - 0.8	56	15	0.1	12	sinking
0.8 - 1.2	56	15	0.1	12	sinking
1.2 - 1.5	56	15	0.1	12	sinking

Starter feeds in particle sizes of 1 mm and 1.8 mm can also be produced as mini pellets. This will result in feed with less dust.

Pre-grow feeds can be purchased as high performance feed with high levels of proteins and fats or as feed with a lower performance containing lower levels of proteins and fats, which will result in less growth of the fish and lower feed cost.

Sizes (mm)	Protein (%)	Fat (%)	Crude fibre (%)	Ash (%)	Pellet
2	50	15	0.7	8.8	floating
2	42	13	2.3	6.4	floating

Out-grow feeds are produced in different ranges of quality. They have been given names, such as 'economy feeds', 'essential feeds', 'superior feeds' etcetera. Out-grow feeds are usually floating but you can still buy sinking feeds which are only pelletized, not extruded. Floating feeds are the most in demand on the market.

Sizes (mm)	Protein (%)	Fat (%)	Crude fibre (%)	Ash (%)	Pellet
3	48	13	1.4	9.6	floating
3	42	13	2.6	6.7	floating
4.5	42	13	2.6	6.7	floating
4.5	37	10	2.4	7.1	floating
6	40	10	2.5	7.2	floating
8	40	10	2.5	7.2	floating

Repro feed is feed for brood stock to optimise the egg development and the survival rate of the fry. 40 mg/kg Astaxanthin has been added to the feed.

APPENDIX 7

Simplified feeding schedule for African catfish based on good quality feed and optimum water quality

- Feeding schedule for juveniles of 10 g growing to 1500 g in 180 days
- Stocking at day 1: 5000 pieces of 10 g
- Feeding in kg

Day	Feed gift	Day	Feed gift	Day	Feed gift	Day	Feed gift	Day	Feed gift	Day	Feed gift
1	2	31	11.5	61	27.5	91	43.75	121	68	151	91
2	2.5	32	12	62	28	92	44.5	122	69	152	92
3	2.6	33	12.5	63	28.5	93	45.25	123	70	153	93
4	2.7	34	13	64	29	94	46	124	71	154	94
5	2.8	35	13.5	65	29.5	95	46.75	125	72	155	95
6	3	36	14	66	30	96	47.5	126	73	156	95
7	3.2	37	14.5	67	30.5	97	48.25	127	74	157	95
8	3.4	38	15	68	31	98	49	128	75	158	95
9	3.6	39	15.5	69	31.5	99	49.25	129	76	159	95
10	3.8	40	16	70	32	100	50	130	77	160	95
11	4	41	16.5	71	32.5	101	50.75	131	78	161	95
12	4.3	42	17	72	33	102	51.5	132	79	162	95
13	4.6	43	17.5	73	33.5	103	52.25	133	80	163	95
14	4.9	44	18	74	34	104	53	134	81	164	95
15	5.2	45	18.5	75	34.5	105	53.75	135	82	165	95
16	5.5	46	19	76	35	106	54.5	136	83	166	95
17	5.8	47	19.5	77	35.5	107	55.25	137	84	167	95
18	6.1	48	20	78	36	108	56	138	85	168	95
19	6.4	49	20.5	79	36.5	109	56.75	139	86	169	95
20	6.7	50	21	80	37	110	57.5	140	87	170	95
21	7.1	51	22.5	81	37.5	111	58.25	141	88	171	95
22	7.5	52	23	82	38	112	59	142	89	172	95
23	7.9	53	23.5	83	38.5	113	60	143	90	173	95
24	8.3	54	34	84	39	114	61	144	90	174	95
25	8.7	55	24.5	85	39.5	115	62	145	90	175	95
26	9.1	56	25	86	40	116	63	146	90	176	95
27	9.5	57	25.5	87	40.75	117	64	147	90	177	95
28	9.9	58	26	88	41.5	118	65	148	90	178	95
29	10.4	59	26.5	89	42.25	119	66	149	90	179	95
30	10.9	60	27	90	43	120	67	150	90	180	95

- Total feed: 8723.65 kg
- Feed conversion rate: 1.25
- Growth: 6978.92 kg
- Total biomass: 6978.92 + 50 = 7028.92 kg
- Harvest: 4700 pieces
- Average weight: 1.5 kg

If you start with a different number of fish you divide the feed gift by 5000 and multiply by the number of fish you have stocked and repeat this calculation for all days. Feeding tables are always an advice and should be a guideline. If the fish cannot eat this amount reduce according to their appetite.

APPENDIX 8

How to hatch brine shrimp (Artemia salina) cysts

(Source: Brine Shrimp Direct)

Introduction and storage

The hatchlings of brine shrimp (Artemia salina) cysts are excellent feed for catfish swim up-fry. They are called naupplii and are of the same growth stage as catfish yolk sac-fry.

Adult shrimps in nature can produce live nauplii immediately, but in adverse circumstances they produce capsulated cysts. When the situation gets better the cysts will hatch, even after a period of many months. The cysts float on the water and can be scooped, cleaned, dried and canned for aquaculture purposes.

Purchased packaged brine shrimp cysts should always be kept under dark, dry and cool conditions. When the cysts are stored for months, it is better to put them in a freezer. Brine shrimp cysts, which are used immediately or in the upcoming weeks, should be separated from the main stock.

Hatching of Artemia cysts

Salt and pH

Artemia cysts start hatching when they are put in a saline solution of 25 g/l; the water should be aerated so that the cysts remain in suspension. Preferably, the salt used for the incubation should be sea salt or industrial salt. Don't use salt which is processed and enriched with iodine and chemicals which avoid clumping of the salt. If the pH of the solution is 8 or below, it should be elevated to a level above 8 with sodium bicarbonate ($NaHCO_3$).

Light

Illumination is necessary to trigger the hatching mechanism within the embryo during the first few hours of incubation. For an optimal hatching rate, it is advisable to leave the light on for the entire incubation period.

Temperature

The incubation time is depending on the water temperature; 26 - 28 °C is optimal. At this temperature hatching of Artemia cysts will take about 24 hours.

Aeration

Aeration is necessary to keep the cysts in suspension and to keep at least 3 mg/l oxygen in the water.

Stocking density of the cysts

The best stocking density for the cysts is 1 g/l. Higher stocking densities result in a lower hatching rate.

Containers

Funnel shaped containers are best for hatching Artemia cysts (Photo 13A).

Photo 13A. A locally made incubator for Artemia cysts.

Harvest and feeding

To harvest the nauplii, remove or turn off the aeration and wait to allow them to separate from their shells. After several minutes the shells will float on the surface while the nauplii will be at the bottom. The bottom of the container has a valve with a fine meshed sieve (see Photo 31, chapter 8). Harvesting can begin now.

Rinse the harvested nauplii with fresh water before feeding them to the catfish swim up-fry, because the hatching medium is dirty and contains a lot of bacteria. The best way forward is to harvest the nauplii directly after hatching, because the quality of freshly hatched Artemia cysts is higher. Surplus nauplii can be frozen and fed as ice cubes.

Fresh Artemia nauplii stay alive in fresh water for hours, so the catfish fry will have a lot of time to consume them.

GLOSSARY

- **Aerobic process:** a biological process which requires oxygen.
- **Amino acids:** the basic units of proteins, which are the building blocks of any living creature. The sequence of the amino acids in the protein determines the type of protein. There are only 23 different amino acids. Some of them are essential in a diet in order to make it complete.
- **Anaerobic process:** a biological process which takes place in the absence of oxygen.
- **Antifungal additives:** additives that inhibit the growth of fungus, which are important to increase the shelf life of products like fish feed. During the rainy season especially, when humidity is high and the environment is favourable for fungus.
- **Arborescent organ:** this organ absorbs oxygen directly from the atmosphere. It consists of cauliflower-like structures situated on the gills supported by cartilage and covered by highly vascularised tissue.
- **Artemia salina:** a species of brine shrimp which can be found in salt lakes all over the world. The shrimps produce eggs which are encapsulated and float on the water. These cysts are harvested, dried, cleaned and sold to fish hatcheries with the purpose of hatching them in salt water and feeding these newly hatched eggs (nauplii) to the fish larval stages.
- **Binary fission:** a reproduction system of singled-cell organisms by one cell splitting into two new cells which creates two new organisms.
- **Biofiltration:** a system which enables degradation of pollutants by bacteria in a bioreactor.
- **Biomass:** at the fish farm this is the mass of living fish.
- **Biosecurity:** procedures and/or measures to protect the fish population at the farm against harmful bacteria, parasites, viruses or biochemical substances.
- **Branchial arches:** gill arches with a bony structure which support the gills.
- **Break-even point:** the point where total cost (expenses) and total sales (revenue) are the same.
- **Broad spectrum antibiotic:** an antibiotic that acts against a wide range of disease-causing bacteria.
- **Blood agar:** a with blood enriched bacterial growth medium. Some bacteria such as streptococci don't grow well if blood is not added to the medium.
- **Cash flow:** payments that bring money into and take money out of a business.
- **Caudal fin:** the tail fin of a fish.
- **Chorion:** the outermost membrane surrounding the embryo.
- **Cryopreservation:** the use of freezing at very low temperatures to preserve structurally intact living cells such as sperm.

- **Coliforms:** rod shaped Gram-negative bacteria. They are a commonly used indicator for the sanitary quality of food and water.
- **Denitrification:** a microbiological process which reduces nitrate to molecular nitrogen under anaerobic conditions.
- **Digestible fraction:** the fraction of feed that can be broken down in the stomach and the intestines, and subsequentially be absorbed by the body.
- **Dorsal fin:** the unpaired fin on the back of a fish which stabilises it against rolling and assists it in sudden turns.
- **Ectotherm:** an animal that is dependent on external sources for body heat.
- **Ectoparasites:** parasites that live on the outside of their host.
- **Endotherm:** an animal that generates its own body heat.
- **Endoparasites:** parasites that live inside their host-animal.
- **Essential nutritional factors:** vital compounds that the body can't make by itself, and are therefore essential for food to prevent diseases and ensure growth and good health.
- **Extruder:** a machine by which a set of mixed ingredients is forced through an opening in a perforated plate or die. In fish feed milling, the pellet produced in an extruder, expands when it leaves the die and causes the feed to float, since air has been incorporated into the pellet.
- **Facultative anaerobic organism:** a single-celled organism which can live with or without oxygen.
- **FCR for fish:** feed conversion rate for fish which is expressed as the amount of feed needed to grow 1 kg of fish.
- **Fisheries:** businesses or industries devoted to catching fish and the science of producing fish for the purpose of providing human food.
- **Gaussian distribution:** the continuous probability distribution shown in a chart. Normally a bell-shaped curve. A higher curve implies that your efforts towards your target are being increasingly met.
- **Gill filaments:** the red, fleshy part of the gills. These filaments absorb the oxygen from the water into the blood.
- **Gill rakers:** bony structures for filtering particles or food out of the water. They don't have a respiratory function.
- **Gonadotropin releasing hormone:** this hormone is responsible for releasing other hormones from the pituitary, such as the follicle stimulating hormone which induces the ovulation of the eggs.
- **Gram-negative bacteria:** these bacteria will not retain the crystal violet stain used in the Gram staining method to differentiate bacteria.

- **Growth curve of fish:** an S-shaped curve which represents the growth pattern of fish and any other living creature. The growth starts slowly, then accelerates into an exponential growth rate, followed by a decline until almost zero.
- **Growth rate expression:** the growth rate of a living creature is always expressed as a % of a unit in a period of time. For the growth rate of fish, it's expressed as a % of the bodyweight in a day. For example, if the fish grow 5% a day, it implies that a fish of 100 g grows 5 g in a day.
- **GSI:** the gonadosomatic index (testes or ovary). The weight of the gonads is divided by the total tissue weight x 100%. For example, a female fish of 5 kg has 500 g of eggs. The GSI of the female fish is: 500/5000 x 100% = 10%.
- **Haemoglobin in fish:** a protein in the red blood cells that transports oxygen from the gills to the rest of the body.
- **Hammer mill:** in a fish feed plant this is a machine used for crushing raw materials, such as maize and soya, into powder by repeated blows of small hammers.
- **Harmattan:** a season on the West African subcontinent, from early December until mid-March, with a dry dusty north-east wind which originates from the Sahara Desert.
- **Hypoxia:** not enough oxygen in the tissues to sustain normal body functions.
- **Invertebrates:** animals that lack a vertebral column or backbone.
- **Locomotory:** ability to move from one place to another.
- **Maintenance feeding level in fish:** the level of feeding which allows the fish neither to grow nor to lose weight.
- **Maxillary barbels:** a single pair of catfish barbels that point sideways from the corners of the mouth. They are the longest barbels of the catfish.
- **Mandibular barbels:** two pairs of catfish barbels pointing downwards.
- **Metabolic fraction:** the fraction of the feed which is responsible for growth.
- **Micronutrients**: the essential vitamins and minerals in the feed that fish cannot produce themselves.
- **Micropyle:** a small opening in the fish egg, through which sperm cells can enter.
- **MSS-222:** tricaine mesylate is a sedation chemical for catfish which is a white powder. At a dosage of 2 g/10 l water, the catfish will be sedated after a few minutes.
- **NAFDAC:** National Agency for food and drug administration and control.
- **Nasal barbels:** a single pair of catfish barbels pointing upwards.
- **Nitrification:** the biological oxidation of ammonia to nitrite followed by oxidation of nitrite to nitrate.
- **Non-specific defence system of fish:** the organs of a fish which protect it against every invader in the same way. This system includes physical barriers, such as the skin and the slime, and the white blood cells which attack any strange invader coming into the body.

- **Non-specific disease signs:** signs which are not specific to a particular disease but are common for many different diseases.
- **Olfactory organ:** the fish's organ for sense of smell.
- **Operculum:** the gill cover.
- **Pectoral fins:** a single pair of fins situated just behind the gills on each side.
- **PHCN:** Power Holding Company Nigeria (before NEPA).
- **Pituitary gland:** a gland that is embedded in the brain and regulates many physiological processes, such as reproduction.
- **RAS:** recirculating aquaculture system.
- **Respiration**: an aerobic or anaerobic chemical reaction in living cells which releases energy.
- **Salmonids:** a fish family widely farmed in cold and moderate climates. Trout and salmon are the most common species within this family.
- **Self-feeder:** a feeding system which allows the fish to eat when they want to, because they can initiate the feed distribution themselves.
- **Septicaemia:** blood poisoning, which occurs when a bacterial infection in one part of the body is spread to other parts in the body via the bloodstream causing multiple infections.
- **Specific defence system of fish:** antibodies that are produced by lymphocytes in response to a hostile invasion. They are specific to the invader or antigen and inactivate it. With a second invasion of the same antigen, the fish's defence system is already prepared and the disease will not break out. Thanks to the antibodies the fish has become immune to this disease.
- **SOP:** standard operating procedure.
- **Standing stock:** at the fish farm this is the quantity or mass of fish which is responsible for the continuous growth of fish to the selling size.
- **SWOT analysis:** technique for understanding strengths and weaknesses and identifying opportunities and treats of a business idea or plan.
- **TAN:** total ammonia nitrogen, which is the total amount of nitrogen in water in the forms of NH_3 and NH_4^+.
- **Trematode:** a fluke which has the ability to attach itself to a host with anchors.
- **Urbanisation:** the shift of a population from farm areas to town areas.
- **Urogenital:** related to both the urinary and the genital organs.
- Ventral fins: a pair of fins which start after the pectoral fins before the urogenital system.
- **Vitellogenesis:** the production of yolk in the egg.

WORD OF THANKS

A word of thanks to all the people who have helped me to realize this book. In particular I want to mention Mr. Steve Elias of S.C.C. Ltd. who motivated me to start writing this book. As well I want to thank my friends Adama Enyojo and Iyodo Dominic for helping me with all the illustrations in the book aiming to make it practical and easily readable, and my wife Mercy for stimulating me to finish this huge task.

Last but not least I want to have a special word of thanks to Heleen van der Maas for her important editing role without changing the context. Certainly she underestimated the work, but never complained. Heleen, it was a real pleasure cooperating with you.

'African catfish farmer's handbook' is a practical handbook written by Willy Fleuren from a background in aquaculture and a wealth of experience in the practical management of cultured fisheries and in different production systems. It is filled with interesting learning points and nuggets of wisdom for a successful enterprise in fish farming. In addition, the author presents his points in the simplest, most understandable, and adaptive form.

Willy Fleuren (58) has been involved in fish farming in Nigeria since 1998. He pioneered the introduction and establishment of the first Recirculation Aquaculture System in Nigeria. A believer in technology transfer, he trained the first set of Nigerians on its use, and they worked directly with him on the early prototypes of the system. This led to an initial explosive wide acceptance of the system through the train-the-trainer approach.

My interest in aquaculture and fisheries is why I took the time to read the manual. It is a handbook that anyone can enjoy. It will be useful for students, start-up and established entrepreneurs, researchers, input suppliers, colleges, universities and government officials. I highly recommend this interesting and educational handbook to all persons interested in fisheries and aquaculture. It is an excellent guide and must-have reference material.

Born in the Netherlands, where Willy Fleuren started fish farming to complete his MSC degree in Aquaculture in 1984, he has been living in Nigeria since 2009. Presently Mr Fleuren lives in Abuja with his wife and children.

Foluke O Areola,
Former Federal Director (Fisheries, Fisheries & Aquaculture, Quarantine)
First Female & Past National President, Fisheries Society of Nigeria (FISON).

Made in United States
Orlando, FL
04 November 2026

90201365R00138